Computational Morphologies

Michela Rossi · Giorgio Buratti
Editors

Computational Morphologies

Design Rules Between Organic Models and Responsive Architecture

Editors
Michela Rossi
School of Design
Politecnico di Milano
Milan
Italy

Giorgio Buratti
School of Design
Politecnico di Milano
Milan
Italy

ISBN 978-3-319-86960-5 ISBN 978-3-319-60919-5 (eBook)
https://doi.org/10.1007/978-3-319-60919-5

Softcover reprint of the hardcover 1st edition 2017

Printed on acid-free paper

This Springer imprint is published by Springer Nature
The registered company is Springer International Publishing AG
The registered company address is: Gewerbestrasse 11, 6330 Cham, Switzerland

Foreword

Founded in 1983, eCAADe—education and research in Computer Aided Architectural Design in Europe—is a nonprofit-making association of institutions and individuals with a common interest in promoting good practice and sharing information in relation to the use of computers in research and education in architecture and related professions. To fulfill this mission, eCAADe organizes an annual international conference and since 2013 additionally a Regional International Symposium. All presented research papers are published in the conference proceedings and archived in the open access platform CumInCAD (http://www.cumincad.org), which holds papers from the worldwide sister organizations ACADIA (North America), CAADRIA (Asia), eCAADe (Europe), SIGraDI (South America), and ASCAAD (Arab region), as well as related conference series.

The eCAADe conference is hosted each year by different universities in Europe, and the conferences constitute a major event where researchers from all over the world can present the latest results of their research and exchange ideas. To help spread eCAADe's goal of fostering high-quality teaching and research in CAAD in new regions as well as in related disciplines, the eCAADe Regional International Symposium was initiated. After the first event 2013 at the University of Porto, Portugal, and the second 2014 at the Bialystok University of Technology, Poland, the third 2015 event took place at the School of Design at the Politecnico di Milano, Italy, on May 14–15.

Extending wider to other professions is also quite new to eCAADe, and we are very pleased that Prof. Michela Rossi and Prof. Maximiliano Romero from the School of Design at the Politecnico di Milano took this challenge. The scientific results shown in this publication in the fields of Design and Shape Grammar, Design and Responsivity, and Digital Heritage clearly widen the view of the eCAADe community. The presented papers show the very high quality of the Italian research, and also some typical, regional focuses that other researchers can learn from.

The eCAADe organization thanks the local organizing committee, particularly Prof. Michela Rossi and Prof. Maximiliano Romero, for their great efforts in organizing this symposium on the theme "Computational Geometries: design rules between nature model and responsive architecture", and for their work on this publication, perfectly summarizing the symposium results.

Joachim Kieferle

José Pinto Duarte
Current and Previous eCAADe Presidents

Preface

Geometry, Rules and Models

The "*Digital Revolution*" is changing each side of our society. In the world of the Industrial Design, which involves as well the renovation of architecture, a deep transformation improved the representation, then the formal reference and eventually the design itself. Facilitating the creation of complex geometries, 3D modeling offered a new, powerful tool, which allowed designers to create easily organic or not regular forms of contemporary international trends. As every actual turning point, the change regarded not only the shape, which means formal features, but just the concept and then the approach to design, namely the design process. The synergy between the digital modeling and the use of prototyping machines can easily transform virtual models into actual objects, anticipating the change in production processes and industrials manufacturing. The production is changing as well. With the 3D printing, the innovation concerns specially the process, leading the way to further changes that on the one hand give new life back to old practices, on the other hand link together the designer and the maker. A new crafts age opens after the crisis that afflicted the industrial age. Actually the world's economic crisis stressed the instability of the global model of the "Western" industrial system, which built its wealth on the production of huge numbers of identical objects. The mass production gave the economic justification for manufacturing production lines to make broadly identical low-cost products, but the industrial model failed with the contemporary decline of consumption. Markets ask for adaptability and product's customization: Digital fabrication answers the requirement of a new economic model that brings back to craftwork. The industrial design was a consequence of the machine age that followed the Industrial Revolution, when Design and production have parted. The industry called a careful project to manage the whole production chain: the design process prior to the making. A common fate linked the industry to design, while it diverged from crafts. In fact, the artisan was able to create objects from a true-scale drawing, without any preliminary draft, because in handwork to make and to design refers to the same actor. Craftsmen just made objects, machines only produce them.

The dichotomy between the creative and the manufacturing phase, which features the industrial process, relegates the contribution of design in the drafting process. The relationship between thinking and making, originally intertwined in the action of the crafts's hand, evolved in design: The drawing played its connection role from making to thinking. It became a technical language instead of being just an active tool in creative process. Today, the integration of parametric generative software with digital prototyping in 3D printing allows the low-cost manufacturing of special pieces from the same code-model and it gets the concept of a single piece married with serial production and it puts again basic design rules at the beginning of the making process. On the other hand, it opens the way for a new type of designer-craftsperson.

A new Industrial Revolution is following this new relationship between design and making, that is new and old at the same time. To manage the transformation of tools without get rid of all the heritage of design culture, we need to re-discover the root of design. It lay in Shape Grammars and in their basic rules, which refers to the growing process of living. The design process becomes the scripting of a code in order to design a product's DNA that starts a morphing process with all its possible variants. Digital technology re-discovers and applies the fundamental principles of formal structure (grammar) and manages how these principles subtend the final form. The digital code subtends the form, and designing comes into play where the first meets the latter. The design merges with the making, they are the work of same actor, they are two steps of the same work.

These developments will have far-reaching implications in design procedures and methodologies. Digital fabrication may offer new opportunities to design, but it needs the development of a different approach in making process. Starting from these awareness, the eCAADe's Regional Workshop in Milan focused on the use of digital technologies and generative software in the design of innovative textures and surface patterns, particularly those which offer high performance to the areas of industrial design, fashion and architecture. The programme involved design and experimentation by the use of algorithmic design and digital fabrication tools, cognizant of ancient principles on the one hand, yet fully aware of innovations in new technologies on the other.

The first aim of the meeting, hosted in one of the biggest design schools in the word, was to focus participants on changes in design and in fabrication, which concern the scholars' attention to the whole process and its transformation through the digital management with the representation of their closed or interlaced systems, as they were organic systems. The second aim was to promote a comparison among schools from different countries and in between Italian schools, which arrived late to digital applications and are conditioned by important architectural heritage that conditions cities developments and attracts main economic and cultural resources. At the beginning of the twentieth century, the industrialization and the consequent social changes led to the birth of design and to radical transformations in architecture. Today, the closed relationship between man and information management makes it possible to amend the object during its production, combining the concept of the single piece with mass production. It is possible to create objects that change gradually their shape. They adapt to different needs of customization, and/or create have responsive element that react to environmental stimulus.

The product and the architecture become changeable and mutant. In the different scales design, from objects to buildings, the morphology changes in order to fit in "*real time*" with the user's needs, following a set of rules that fix the relationship between changing parameters. The parametric design is a digital structure of dynamic links in between different factors: generating events (input), the project (digital actions' processing of shape), and formal representation (output). It is a way to create responsive objects and architecture that are able to react to externals stimulus.

Parametric representation includes the variable "time", and it concerns the simulation of movement in order to check the handling of responsive elements. Digital design goes over the virtual representation and improves the interaction between the designer and the whole environment. The effect of parametric design and rapid prototyping on the development of new forms of kinematic surfaces and responsive architecture is flagrant. This digital evolution is powered by a renewed relationship between man and computer, or more generally between the human being and the digital space. The key element of the innovation is the centrality of computing in the design process, such as in new products development. Computing is the intelligent soul of design.

An articulated design process subtends the representation of this complexity. It put together on parametric modeling, on kinematic simulation of models, on digital prototyping, and on the digital acquisition of experimental mock-up models.

This book gathers several contributions to the workshop: some introductory lectures and several personal researches. Speeches are organized in three parts:

The first section focuses on design models and shape grammar, up to the management of complex forms. It begins with an analysis of the organic model and the basic theory of design.

Innovative concepts derive from organic models through the use of generative design tools and digital fabrication technologies. Shell, lattice, and grid structures can underpin patterns and surface textures. The knowledge of the Nature's harmonic relationships can be traced to Pythagorean doctrine, Vitruvian principles, and later Renaissance thinkers.

In the nineteenth and twentieth centuries, the knowledge of biological systems development provided inspiration to theoreticians like Theodore Cook, D'Arcy Wentworth Thomson, Ernst Haekel, Jay Hambidge, Matila Ghyka and Keith Ctitchlow, and practitioners like Richard Buckminister Fuller and Charles-Édouard Jeanneret-Gris (better known as Le Corbusier). In contemporary times, digital technologies allow an easy access to patterns and textures, as well as knowledge of proportion and balance, at the nano-, micro- and macro-levels; such technologies offer sources of historical inspiration in the fields of the visual arts in general and industrial design and architecture in particular. In the Nature's models, the industrial design and the architecture find their common origin in ancient concepts of basic design.

The second part concerns the relationship between Design, Architecture, and Responsivity. The term "Responsivity" is used in relation to the changing morphologies of the architectural artifact. As early as the 1970s, visionary thinkers like Nicholas Negroponte proposed that advances in artificial intelligence and the miniaturization of components would soon give rise to buildings, capable of changes in the external and internal environment. Thanks to technological advances, today we are able to manage these changes through pre-programmed mechanism of real-time response and feedback embedded in inhabitable spaces. It is possible to consider the building as a system which adapts its behavior to information acquired about its users. Information external to the building could also be integrated into the process, so as to respond to a multitude of condition able to create new forms of experience and expression.

The third section deals with digital heritage. "*Heritage*" is a broad term that refers to the study of human activity not only through the recovery of remains, but also through tradition, art, and cultural evidences and narratives. "*Digital heritage*" is an active area of research throughout the last decade, and it is a process of research and transmission of ideas and values and a knowledge that includes the material, the intangible and the virtual.

With the advancement of technology, digital heritage projects have enhanced their capability to increase interaction and manipulation of the research object, facilitating new findings. For example, augmented reality, the integration of digital information with the user's environment in real time, seems to be of particular relevance in the fields of archaeology, architecture, art, and city planning.

Furthermore, a few recent cases demonstrate the possibilities of heritage interpretation through use of modeling generative tools. These researches indicate new frontier in study of cultural heritage, while opening up possibilities of revealing complexity levels that could not be managed only a few years ago.

That demonstrates how digital heritage interpretation can be considered as a new process and at the same time how to ensure multiplicity in understanding the past.

That focus on cultural heritage expresses a special feature of Italian schools that apply innovative technology in the development of a new approach to the management of material and immaterial values. Actually, the application of digital technology to cultural heritage is an important topic in Italy, not only concerning digital representation, augmented reality, virtual museums, and heritage conservation. It may be an useful research tool that links tradition and innovation. New technological inventions have always made it possible to investigate and improve our knowledge of the world—think of the introduction of perspective and the use of the frame as the basis for Renaissance architecture and painting or of the studies of the lens by Galileo as the basis for the telescope and microscope. These tools, made thanks to scientific research, amplify human capabilities, while increasing the capacity for investigation and analysis and thus enabling new discoveries. Similarly, the research covered in this publication can be considered as an extension of human thought, enabling the understanding of formal properties and complex nonlinear phenomena which cannot be managed with traditional tools. In a word where computer-assisted design accompanying the designer from the generation of the form through its digital fabrication, the integration with theoretical analysis and comprehension tools able to maintain a high level of coherence is increasingly necessary. The creation of conceptual devices able to define correct methodological procedures represents the intellectual challenge of the future for the design disciplines.

Milan, Italy

Michela Rossi
Giorgio Buratti

Contents

Architectural Fabrication: Towards Eco-Digital Design to Build Process in Architecture

Paolo Cascone, Elena Ciancio, Flavio Galdi and Andrea Giglio

Abstract The article focuses on the applied researches and built projects that I have developed with the COdesignLab in the field of architectural fabrication. Such work investigates on the role of parametric design and digital processes for building performative shells. As Mohsen Mostafavi pointed out in his book "*Ecological Urbanism*" (Menges et al. in Advancing wood architecture: computational approach, Routledge, London, 2016) the fragility of our planet could be an opportunity for "*speculative design innovations*" rather than "*technical legitimation for conventional solutions*". With this cultural premise, as mentioned by Saskia Sassen in her essay "*Global city 20 years later*" (Cheret and Schwaner in Urbaner Holzbau. Handbuch und Planungshilfe, A. Seidel, 2013), the research project considers global cities as spaces for "advanced knowledge products". The advanced production processes, as described by Chris Anderson in his book "Makers" (Menges in Performative wood: integral computational design for timber constructions, Acadia, 2013), are driving the—so called—"New Industrial Revolution"; thanks to the digital fabrication technologies the traditional relation between designer, producer and consumer is rapidly changing. Meanwhile the open source technologies are already affecting the economy supporting the development of low-cost computer aided manufacturing (CAM) machines creating new economic and social opportunities integrating traditional manufacturing processes for customized solutions. Therefore the applied research responds to the increasing

P. Cascone (✉) · E. Ciancio · F. Galdi · A. Giglio
COdesignLab, Naples, Italy
e-mail: paolo@codesignlab.org

E. Ciancio
e-mail: info@codesignlab.org

F. Galdi
e-mail: info@codesignlab.org

A. Giglio
e-mail: info@codesignlab.org

M. Rossi and G. Buratti (eds.), *Computational Morphologies*,
https://doi.org/10.1007/978-3-319-60919-5_1

desertification of our planet connecting digital fabrication processes with natural materials and smart construction techniques in order to evolve the vernacular architecture as classified by Bernard Rudofsky and Paul Oliver. As a matter of facts building with wood and clay/ceramics has many desirable properties as a building material: thermal mass characteristics (energy efficiency), humidity controlling properties (environmental control), plasticity of form (structural stability). For the wooden structures the inspiration was coming from Japanese traditional construction techniques of interlocking systems revisited with the aim of using local material according to climatic conditions. In order to develop customized and performative solutions we have integrated subtractive technologies using CNC machines and laser cutters. In the case of woodless constructions we have been inspired by the sub-Saharan spontaneous earth-made architecture and by the work of Hassan Fathy and Fabrizio Carola in Mali. The projects develop a catalogue of vaults and domes as a performative tectonic systems for architectural dwellings. Domes are realized using 3D printing technologies extruding local clay. Bricks are designed with a parametric approach and assembly bridging traditional techniques and CAM machines. After the first scale 1/1 prototypes built in Africa and Europe in the last years we have developed scientific protocols for the architectural fabrication within the possibility of self-producing and self-assembling structural components. Such scientific work has been presented in year 2015 at the IASS International Conference (structural design) and the PLEA International conference (passive and low energy architecture). This with aim to create a new discipline able to provide both structural efficient and anti-seismic solutions as well as energy saving performances for sustainable living. In year 2017 we will use the results of such applied research in order to design and self-build the African Fabbers School in Camerun.

Keywords Wood · Clay parametric design · Performative shells · Digital fabrication · Open source technologies · Earth made construction · Traditional techniques · Computer-aided-manufacturing · Anti-seismic optimization

1 Wooden Systems

Wood "*embodies a rich history and has deep cultural roots, while at the same time providing striking prospects for the future built environment, which mainly stem from the confluence of two important factors: the ever-expanding design, simulation and fabrication possibilities enabled by computation on the one hand, and wood's virtues as one of the few ecologically sound building materials on the other*" [1].

For many centuries wood has been the single most important building resource. Until the end of the eighteenth century, more than 80% of all buildings were timber structure [2]. The structural differences in the raw material were employed for

Fig. 1 AA Hooke Park. Wood chip barn

specific building elements through manual wood-working techniques. These fabrication methods allowed to precisely cut and bend the harvest log following the natural grain direction. As a result, the craft-based design and construction techniques were in close relation to the available material and its specific, highly anisotropic characteristics (Fig. 1).

Today the emerging digital technologies (like CNC machines and milling or robotic arm) and Computational design can be an opportunity to evolve and improve these craft based techniques towards approaches that let employ complex behavior rather than just modeling a particular shape or form. Moreover these tools are putting capability to produce and control physical products into the hands of everyone (Wikihouse, 2010). The transition from currently predominant modes of Computer Aided Design (CAD) to Computational Design allows for a significant change of employing the computer's capacity to instrumentalize wood's complex behavior in the design process. In Computational Design, form is not defined through a sequence of drawing or modeling procedures but generated through parametric, rule-based processes. The ensuing externalization of the interrelation between algorithmic processing of information and resultant form generation permits the systematic distinction between process, information and form (The "wood chip barn" project of AA Hooke park can be an example). Hence, any specific

Fig. 2 Wikihouse prototype

shape can be understood as resulting from the interaction of system-intrinsic information and external influences within a morphogenetic process [3]. The following projects are the results of a research that goes in the this direction (Fig. 2).

1.1 Grid(h)ome Pavillion

Design concept: inspired by the Frei Otto's researches and projects about gridshells and light-weight structures, the project creates a relational space for the garden of the Casa dell'Architettura of Rome. The structure has been designed in collaboration with CMMKM and realized during a two weeks workshop organized by the Italian Institute of Architecture.
Design driver: daylight control, passive cooling ventilation
structural system: gridshell
material system: wood, recycled plastic panels (Fig. 3)

Fig. 3 Grid(h)ome pavillion. Detail of structure and installation (*below*)

1.2 Smart Shell

Design concept: smart shell is designed as a transportable, fast-deployment and energetically self-sufficient structure. The design process investigates the relation between high-tech design and low-tech construction.
The project provides flexible solutions for different uses from low-cost dwelling to off-grid emergency hospital etc.; the structural system consists in foldable modular arches made of wood covered by photovoltaic textile strips. Smart shell has been selected by UN-Habitat to participate at the 6th World Urban Forum
Design driver: thermal comfort, solar energy
structural system: interlocked arches
material system: wood, photovoltaic textile (Fig. 4)

Fig. 4 Smart Shell, 6th World Urban Forum

1.3 Folding Foyer

Design concept: after the fire of the Naples Science Centre, the foyer of the Galileo 101 theatre was damaged. Therefore we decided to develop a temporary solution with a fast deployment wooden structure repurposing a container box.
The project is the result of a collaborative process involving engineering and architecture students. The structural system consists of a continuous roof which integrates digital manufactured shading cinematic panels.
Design driver: light
structural system: folding structure
material system: wood (Fig. 5)

Fig. 5 Folding foyer. Details of the continuous roof and CNC machinary production process

1.4 Open Air Lab

Design concept: the project is conceived as a small infrastructure realized in a public space of Dakar providing an open air laboratory for local craftsmen. The construction of the structure has been realized in the framework african Fabbers project through a participatory process involving the local community and european makers with the aim to support the creation of the first Fablab in Senegal. The project has been selected to participate at the Dakar biennale in 2014.
Environmental driver: rainwater harvesting, thermal comfort, solar energy
structural morphology: branching system
material system: local wood, bamboo and photovoltaics (Fig. 6)

Fig. 6 Open air lab, construction process (*below*)

2 Earth Systems

Earth might be considered as one of the oldest building material, due to its large diffusion and shortness of time from deposit to treatment. As a matter of facts, building with ceramics has many desirable properties: thermal mass characteristics (energy efficiency), humidity controlling properties (environmental control), plasticity of form (structural stability). We can find a modern point of view about the revaluation of using earth materials in Fabrizio Carola's project in Mauritian. Concrete was dismissed as too expensive and unsuitable, and wood excluded because this area suffered from desertification, hence Carola was left with the availability of earth and stone which were utilized to make components for dome-like structures [4].

The potentials of using this material can be explored in the field of the advanced design trough the computational processes, setting cause-effect relationship between form and performance fitting the specific inputs of a construction.

Fig. 7 Quake Column-Emerging Objects

Fig. 8 Desert clay, COdesignLab (Marrakech)

The digital outputs of form-finding operations, match to the need of using advanced digital fabrication technologies for production (CAD/CAM processes).

The most appropriate digital fabrication technique for earth, as also described by Emerging Object studio, co-founded by the associate professor Ronal Rael of the Berkley University and Virginia San Fratello, is that of 3D printing with semi-fluid materials. Trough LDM (liquid deposition manufacturing) it is possible to construct a three-dimensional element by the addition of successive layers of clay compounds

Fig. 9 Kaedi Hospital, Fabrizio Carola

which solidify during the deposition step. The research on earth material and digital processes has been carried out by COdesignLab/Paolo Cascone in different parts of the world trough professional experiences and practical applications which are set out below as paradigmatic cases (Figs. 7, 8, 9).

2.1 Earth[Play] Ground

Design concept: the project was developed with the atelier Paolo Cascone (ESA-Paris) in collaboration with the local community of Sourgoubila.
The space allows the children of the village to play and study in an overshadowed playground.
The design process intended to develop high-tech design/low-tech construction approach to performative design merging computational process within vernacular construction techniques from Sahel region. After a research by design workshop developed in Paris based on the use of earth bricks for vaults we have discussed with the village the project of building a primary school playground (Fig. 10).

Fig. 10 Earth[play]ground, atelier Paolo Cascone (ESA-Paris)

2.2 *Cultural Centre*

Design concept: the project is the result of the collaboration between the Aga-Khan awarded architect Fabrizio Carola and Paolo Cascone in order to design a New Cultural Centre in Sevaré.

Therefore the structure design process aims to develop an innovative approach to build dome morphologies.

The cluster of diversified domes is generated through a parametric process considering functional and spatial needs achieving environmental performances (Fig. 11).

Fig. 11 New Cultural Centre (Sevaré)

Such project investigates on possible innovations of vernacular architecture integrating advanced technologies with traditional construction systems and local earth as material.

2.3 Programmed Matter

Design concept: the environmental installation is a tridimensional structure of 2 per 2.5 m, floating over the entrance hall of the academy; it's composed of a volumetric pattern of about 100 ceramic pieces and its texture is developed through a generative algorithm. Components are fabricated in three different technics: 3D printing with clay, lathe and wick. In this way we established a dialectic interaction between the innovative approach of Paolo Cascone, the artistic knowledge of the professors of the academy (Giancarlo Lepore, Rocco Natale and Luisa Valentini) and the traditional technic of the ceramic masters (Marcello Pucci and Orazio Bindelli) that have cooperated to the realization of this collaborative project. The pattern is generated by the assembly of self-similar components inspired by bio-mimetic structures (diatomen).

Fig. 12 Programmed matter is a tridimensional structure, floating over the entrance hall of the academy, (*below*) 3D printing with clay

The configuration derives from a "*frozen movement*" that can influence the ambient by filtering the light and amplifying the acoustic sources. Indexical floor is a paper surface that graphically shows the morphogenesis of every component by numeric series. the texture that bears and links the single elements of the installation physically and metaphorically represent the relationship between knowledge and technic that inform the work of Paolo Cascone.
Programmed matter, first collaborative work developed in an academic framework in Italy following an innovative methodologic approach, suggests the introduction of new knowledge in the high artistic teaching (Fig. 12).

2.4 Ceramica Performativa

Design concept: the project explores in an innovative way the relation between computational design, digital fabrication and natural material systems for a performative architecture.
Taking inspiration from the Paolo Soleri project (www.arcosanti.org) of Solimene's ceramics in Vietri (Italy) where the prototypes have been realized. Ceramica performativa aims to develop an innovative concept of high/low tech factory for ecological design and architecture.
The exhibited prototypes are part of an applied research on architectural fabrication and ceramic structural skins.
The structural system generates a porous wall able to modulate the daylight an the natural ventilation.
The inner cavities are designed for the water circulation creating an evaporative cooling dynamic in order to improve the thermal comfort performances (Fig. 13).

2.5 Hacking Gomorra/Advanced Furniture

Design concept: the work exhibited is part of an applied research developed by Paolo Cascone on the role of advanced technologies for the intervention of urban self-re-generation.
The project proposes an innovative approach to the transformation of the so called *Vele of Scampia* (Naples) through a process of collaborative hacking of the existing building. The structural components conceived for the project of architectural fabrication of the vela become design artefacts (lamps, tables, vessels, shelves exhibited in the gallery) produced through a process of 3D printing using natural and recycled materials.

Fig. 13 Ceramica performativa

The project aims to develop, thanks to the dialogue with Ugo la Pietra, a critical reflection on the notion of productive cities and on possible application of digital fabrication technologies for contemporary product design and architecture (Fig. 14).

Fig. 14 Ceramica performativa

3 Conclusion

The above mentioned applied research methodology will be the corpus of the African Fabbers school, the first school of urban ecologies, self-construction and digital fabrication in Africa.

The school project, after the participation at the Dakar and Marrakech Biennale, has been selected by the Italian ministry of foreign affairs to be implemented in 2017 in Cameroon in partnership with the COE NGO. The research by design agenda and the educational programme, is based on the idea of bridging the african and the european designers clusters through community oriented projects and applied researches. therefore the AFS will be part of an international network of design schools and research laboratories. The project aims to respond to the lack of design school in the region in order to investigate on the growth of African cities putting emphasis on an ecological agenda: exploring the interaction between african material systems and computer aided manufacturing technologies, developing performative prototypes through an advanced craftsmanship approach for sustainable living. The school itself will be designed and realized within a collaborative approach creating an innovative way of rethinking post-vernacular design in relation with advanced technologies ad performative tectonics.

Credits

Grid(h)ome pavillion

site: Rome (Italy)

team design: Paolo Cascone (architect) with Sofia Colabella, Bianca Parenti, Sergio Pone (architect)

team construction: P. Cascone, S. Colabella, B. Parenti, S. Pone, A. Fiore, B. d'Amico, I. Lupica, D. Lancia, S. Tapsoba with the workshop students

partners: Selve del balzo, CMMKM.

Smart shell

site: Naples (Italy)

client: STRESS consortium

team: Paolo Pascone (architect) with Elisabetta Corvino, Pietro Nunziante, Silvia Piccione (environmental analysis)

team construction: Paolo Pascone (architect), Elisabetta Corvino, Elena Ciancio, Imma Polito, Andrea Giglio, Giuliano Galluccio, Delia Evangelista, Silvia Piccione,

Rossella Siani, Claudia Balestra, Achille Piroli

technical partner: Selve del balzo

photovoltaic: aurora, nhp neaheliopolis.

Folding foyer

site: Naples (Italy)

client: le Nuvole

team: Paolo Cascone (architect) with Elena Ciancio, Flavio Galdi, Giuliano Galluccio, Andrea Giglio, Imma Polito

workshop students: Emma di Lauro, Federico Forestiero, Luigi Gentile, Tullio Grasso, Giorgio Lauro, Gianluca Montone, Laura Moscarella, Matteo Nativo, Marco

Naclerio, Alessandra Stefanelli, Agostino Viglione

technical partner: Urban FabLab

Partners: ARUP, Selve del Balzo, Roland, Sintesi Sud, Le Nuvole, DiST, DiARC.

Open air lab

site: Dakar (Senegal)
client: Ker Thiossane - Defko ak Niep lab
team: Paolo Cascone (architect) with Elena Ciancio, Flavio Galdi, Giuliano Galluccio, Andrea Giglio, Imma Polito
echnical partner: Urban Fablab
partner: Fondazione Idis - Città della Scienza, Fondazione Architetti e Ingegneri liberi professionisti iscritti Inarcassa, Ker thiossane, Dakar biennalE.
local community: Roland Assilevi, Ahmadou ba, Biggiie, Demba Camawa, Ibra Cassis, Moustapha Coly, Seydou Dièmè, Stefano Ferrero, Racine Gaye, Alessia Guardo, Ariss Godwill Hounkpatin, Dodji Koffihonọu, Daouda Kote, Alexandre Lette, Nolèye Manè, Mariana Michalcikova, Susanna Molina, Modou Ngom, Angelina
Nwachukwu, Désiré Nwaobasi, Mame Less Jean Seck, Momar Sylla, Ker Thiossane, Marion louisgrand, Momar François Sylla
thanks to: Madre Museum Naples, Afropixel Dakar

Earth[play]ground

site: Sourgoubila (Burkina Faso)
client: Sourgoubila Primary School
team: Paolo Cascone (architect) with Pascal Angel, Geoffroy Griveaud, Christophe Lefevre, Matthieu de Lacvicier, Ali Benabdellah, Ayda Bennani-Smires, Edouard
Fenet, Mathieu Quilici, Fatou Dieng, Youssef Haddadi, Victor du Peloux, Nadège Guion, Océane Patole, Murielle Teguel, Omar Ghaiti, Juliette Rubel, loubna Touzani,
Marguerite Bureau, Nina Rotter, Salwa Essoussi
construction team: Atelier Paolo Cascone with local community
consultant: la Voute Nubienne
partner: Esa.

Cultural centre

site: Sevarè (Mali)
client: NEA
team: Paolo Cascone (architect) with Fabrizio Carola (architect), Stephane Tapsoba, Helena Agurruca.

Programmed matter

site: Urbino (Italy)
client: Urbino Academy of Fine Arts
team: Programmed Matter is a collective project developed by paolo cascone (architect) with Flavio Galdi, Imma Polito, Elia Amaducci, Li Guoyuan, Huang Lidong,

Ambra Lorito, Noa Pane, Valeria Pilotti, Marika Ricchi, Jessica Pelucchini and Giancaelo Lepore;

technical partner: Codesignlab, Urbanfablab, Spazio geco, Keramos.

Ceramica performativa

site: Milan (Italy)

client: the XXI International exhibition of the triennale di Milano

team: Paolo Cascone (architect) with Andrea Giglio, Elena Ciancio, Flavio Galdi

technical partner: 3d Italy, Solimene ceramiche, Urban FabLab.

Hacking gomorra // advanced furniture

site: Milan (Italy)

client: Milan design week 2016

team: Paolo Cascone (architect) with Elena Ciancio, Andrea Giglio, Imma Polito and Flavio Galdi

special thanks to 3DItaly and Solimene Ceramiche.

References

1. Menges, A., Schwinn, T., & Krieg, O. D. (2016). *Advancing wood architecture: Computational approach*. London: Routledge.
2. Cheret, P., & Schwaner, K. (2013). *Urbaner Holzbau. Handbuch und Planungshilfe*. A. Seidel. Berlin.
3. Menges, A. (2013). *Performative wood: Integral computational design for timber constructions*. Milan: Acadia.
4. Vargas, D., & Cascone, P. (2010). *The legendary Fabrizio carola: "Respect for nature, for the country and for its people is a part of me. Respect stops me from abusing my freedom"*. Domus n° 940, La nuova utopia.
5. Airsoldi, S. (2014). *Urban [FabLab] ecologies*, Ottagono n 247.
6. Anderson, C. (2012). *Makers. The new industrial revolution*. USA: Crown Publishing.
7. Rudofsky, B. (1964). *Architecture without architects*. New Mexico: University of New Mexico.

Part I
Form and Code

On Rules and Roots: The Organic Model in Design

Michela Rossi

Abstract Digital tools and new technology are changing all aspects of today life, up to architecture and design concepts. They allow a new approach to the nature as a model, which is a concept that drives the theory of Architecture trough centuries. Then the affirmation of digital technologies is radically changing the application of organic models, which does not concern the shape any more, but just the morphing process. The aim of these notes is to stress significance, reasons and evolution of this reference that confirms the strong relationship between the theoretical heritage of architecture and its today developments, due to the application of digital technology to design, fabrication and construction. The making process find its theoretical validation in shape grammars and in traditional concepts of module, tessellation and pattern, which rules the design of decoration, surfaces and ordered conformations. Digital application of old basic design's confirms the value of this heritage as constant of architecture theory.

Keywords Pattern · Tessellation · Lattice · Symmetry and asymmetry · Organic models · Basic design · Design theory

1 Introduction

Contemporary architecture has always set the concept of order. It evokes the continuous stress between the tidy cosmos of creation and the chaos that preceded it, with surfaces, which suggest continuous transformations. Its shapes are inspired by the earliest studies in topology, whose spaces no longer obey Cartesian rigor, and imply a continuous formal evolution. The reference to the mathematics often suggested new ideas to architecture and still does, not only because of their

M. Rossi (✉)
School of Design, Politecnico di Milano, Milan, Italy
e-mail: michela.rossi@polimi.it

M. Rossi and G. Buratti (eds.), *Computational Morphologies*,
https://doi.org/10.1007/978-3-319-60919-5_2

applications in computer graphics but mostly because of the suggestion of innovative concepts, the same way that the "other" geometries that negated Euclidean geometry's postulates did. Walls and roofing bend, twist, merge in complex surfaces, without any apparent reference to their previous compositional rules, which were the goal of static, functional and esthetic needs, according to Vitruvio's triad *firmitas, utilitas, venustas.*

Computer-aided design is able to apply complex mathematics and architecture develops new shapes with bending surface—such as nurbs—and complex forms, using of the random repetition of simple algorithms, sometimes without the designer even being aware of the mathematics behind it, thanks to software that allows to model shapes directly. Thanks to a controlled randomness, the computer introduces chaotic elements into the architecture layout, imitating the contemporary world. In its articulated morphology it is difficult to recognize a classical concept of order, but a rule does exist and it is a code that manages the shape. Nevertheless, architecture still requires a balance between inside and outside forces its elements, therefore it cannot avoid the enforcement of rules that ensure harmony of shapes and strength. Design simply develops a different answer and then shows new forms, because of a different approach to models man always found in the organic reference of nature.

The design making process follows an organic-wise growing concept and the observation of nature offers a valid approach to design teaching. Drawing is the essential language of architecture. It expressed concepts through simple signs that are correlated to actual shapes of objects. A geometric thought ruled design and its representation. In the full digital age, the representation of virtual model is no longer the project media only: digital design is the goal of the development of a process that applies transformation rules to a ordered system. The design overlaps the making process through the same digital model. The last is sufficient in itself to the "making"; therefore digital fabrication eliminates the gap between the designer and the worker.

The scripting is the new drawing, because it manages shape grammars and objects morphology. This fact "brings back" to the old world of craft, when the drawing was not a mediated instrument of the project's language but it served directly the construction.

The man's inspiration from natural forms is a recurring citation in treatises and in historical literature, as well as in recent theories, which incorporate many concepts of basic design. The statement that nature is the first model of design has guided the theory of Architecture for centuries, from Semper and on. Along times, it became a myth that showed different evidence. Natural forms appeared first in ornaments and decoration, then they inspired building structures, later the organic model becomes a cyclic recurrence in Architecture design. What changes is only "how" man apply nature's models to its creations. In facts the late development of digital technologies allows a "leap" in the organic reference of design, which does not refers to the form as a static element but to the growth as a dynamic process of transformation. This improves the evidence of organic forms in design objects and therefore it increases the importance in design learning.

This paper resumes a wider research on the role of organic models in design teaching. In the starting phase, the aim was to retrace the significance of organic models in Architecture as a teaching tool in drawing classes. Before digital design became a reality, the goal was simply to understand the rules of form and growth to master shape grammar in design. Today it is the way to master the scripting of generative application as a rational process, avoiding that the organic reference became a bare formalism. That previous didactic experience that acquires new significance with the development of digital technologies, to introduce the application of design rules as a dynamic process, thanks to generative software. It becomes basic information for a competent application to digital design. It states the historical roots of digital design and demonstrates the timeless importance of few basic rules of shape grammars (Fig. 1).

The first research aimed to explain why and how nature became the best design model. This paper focuses on the transformation of the reference to natural models along centuries, depending on development of scientific knowledge. The first theoretical knowledge came just from the ancient overview of tidy Geometry in nature,

Fig. 1 Ernst Heckel (1904) *Kunstformen der Natur,* tables 1, 9, 29, 41, 69, 91

founding the reasons of Geometry as a creation rule, explaining growth process laws in living beings and therefore their imitation in architecture, eventually the evolution of the organic approaches in design, which started from the imitation of form to reach the imitation of life. The paper pursuit different aims:

- to explain the origin, the evolution of nature's myth;
- to understand the scientific reason of nature's configurations;
- to show the evidence of basic shapes and patterns in the organic models and their imitation in design;
- to help the management of shapes as a morphology process.

2 Nature as a Design Model

The perception of beauty was (and perhaps still is) related to a net order that rules the repetition of basic self-connected elements in different conformations. As a matter of fact, nature experimented with several variations on the same theme, combining simplicity and complexity [1].

The eurythmic conformations of nature's creations inspired mankind since antiquity. Men admired and copied their mathematical regularity in the design of ornaments. Regular patterns are evident in the shape of (lower) organisms, such as shells and flowers, becoming the first source of imitation. It was found in the chemical structure of the matter, as well.

The theory of Architecture used the reference to nature as a organic rule in the concept of complex artefacts, because it combines functionality, durability and aesthetic value in a sole expression. The imitation appears in structural design with aesthetic and formal values. The classical theory of the Order is exemplary but any human inventions develop from, or refer to, nature's example.

The imitation is linked to knowledge, because it is the first, elementary way of learning; in facts the man searched and found the inspiration to solve any constructive problem in the nature, which offers a great variety of formal solutions. Its regular configurations became a hard reference both as aesthetic and as structural model. They inspired the general layout too. Furthermore they invited to the search of an absolute law of harmony that pushed to the empirical birth of Science. So the wonder for nature characterized the birth of Western thought. It taught Architecture and led to knowledge, documenting how the man linked the formal symmetry of the natural balance, to the divine perfection.

Classical philosophers sought the answer to the origin of things and life in the observation of the physical world, trying to explain the law that rules the order of concatenated conformations. Heraclitus said that to achieve the wisdom you need to understand the principle that is at the origin of things, namely the $\lambda o \gamma o \sigma$ = *logos* [2].

This concept goes through Greek philosophy, for which things are made of numbers and geometry, with the meaning of word and speech, but also of measure.

The concept of proportion, apparently based only on aesthetic principles, before Galileo's innovations has been the first reference in structural design. The statement of a link between structures stability and visual harmony stresses the evidence of the Vitruvian triad [3].

Several scholars of Modern Age searched for a scientific explication of regular form in organic models as an answer to design questions. Kepler and Cassini studied the regular forms of bees' structures [4]. Galileo, who is the father of modern science, said that the book of nature is written in the characters of geometry. Descartes reconciled numbers and shapes and Kant stated that just nature can introduce Mathematics into natural philosophy. Later Einstein himself expressed the same concept, translating the link between energy and matter in mathematical terms.

Even without any final answer to the secrets of the universe, man found the solution to design and aesthetic problems in the observation of creation. If the Vitruvius statement "*deus architectus mundi, architectus secundus deus*" is true, in order to imitate God, man should really just imitate nature.

To imitate does not mean *to copy*. It means to reinvent, namely to understand and transform. Thus the imitation requires a deep knowledge, which follows a careful observation. Leonardo's work is a good example. The imitation of nature in design emphasizes the significance of observation and the role of drawing in the understanding of formal laws, which requires the management of shape grammars trough geometrical rules. Kepler's statement that Geometry is an absolute concept, it requires knowledge of formal phenomena of nature, which were the first basic assumptions of empiricism [5] and then of the experimental science (Fig. 2).

The acknowledgement of mathematical rules in nature is the focus of design imitation. It supports and justifies the application to design.

In nature's realms there are all forms of geometry, from simplest, such as regular polygons in the circle, to the more complex ones such as logarithmic spirals and minimal surfaces, as well as a wide variety of patterns and lattices with complex symmetric aggregations until the articulated forms of higher organisms. All these

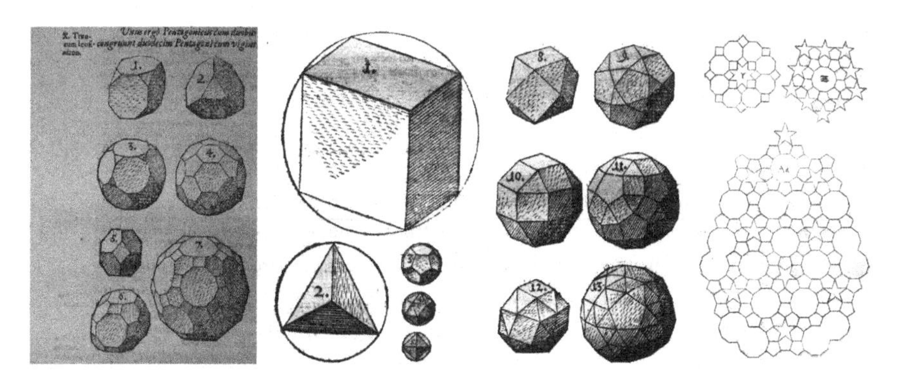

Fig. 2 Kepler's study of regular lattices in regular and semiregular polyhedron and complex plane tessellations

forms are never casual [6]. The geometric structure of molecules builds spatial lattices that are similar to those of Fuller's reticular structures. When it gets larger it turns regular forms into crystals, such as snow crystal.

A regular pattern affects the morphology of living organisms, for which the shape is an element of recognition and survival, and constitutes the final goal of the process that is implicit in life. This fact, which is evident in the net geometry of lower organisms, recalls the concept of Alberti's concinnitas. The last is the fundamental nature's rules that lay in perfection of Geometry.

In complex organisms the form is ruled by the gravity and the growth is regular but not homogeneous. It develops applying logarithmic relationships, with solutions of increasing complexity up to the mechanical efficiency of the vertebrates' skeleton.

3 The Nature's Law: Symmetry and Asymmetries

Alike builders, the nature must operate with materials of predetermined magnitude. The form is defined by the absolute or relative size properties in any directions and it is subject to the chemical and physical laws of matter. D'Arcy Thompson explains that organic forms are the effect of the physical system, which obey to pressure balance in cells and to mechanical forces in larger structures. Then the force of gravity controls the shape of higher organisms, while the surface tension determines the cells' shape that tends to the sphere and related forms. The pure, natural form gains the lowest surface according to the internal volume. External forces produce transformations in the shape to balance the system according to the criterion of maximum efficiency that characterizes the nature (Fig. 3).

Cells have a similar behaviour with the soap bubbles and they reproduce the Plateau's surfaces, whose symmetries appear in configurations of several protozoa [7]. *Radiolaria*, which are single-celled organisms with a silicate skeleton, show the full set of regular polyhedron. *Foraminifera* have a wide variety of allied forms,

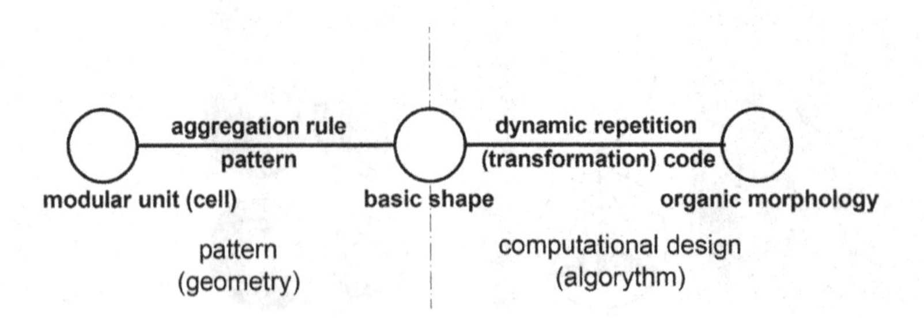

Fig. 3 Organic model and computational design: the development of design from "shape composition" to "form generation"

which document a continuous transformation between related structures that is not due to evolution [8].

The body is more than the simple sum of its members: it is the $\tau\varepsilon\lambda o\sigma$ = *telos*, which is the final goal of the living process of generation and growth that shapes its own form. Natural structures develop according to their final purpose and their body must be compatible with life, which relates to reproduction that is the basis of life. The primary element of biology is the cell, a living aggregation of organic molecules that are in turn compounds of chemical elements. The morphology of living organisms expresses the maximum efficiency of the biological process according to environmental conditions and the exigencies of cells' reproduction in the continuity of life. The form of organisms is the result of their growth from a single beginning cell that divides itself according to plans articulated in an increasingly complexity.

The growth is typical of life, but it is not its prerogative because it appears in crystals too. The transition from inanimate world to life is a unaccounted issue, but the origin of life seems to be due to a formal fact: the symmetry breaking. In fact symmetry is the main factor of stability that means absence of movement. On the contrary the asymmetry's loss of balance involves the movement and then a change. Life is a continuous changing.

Indeed the production of exclusively asymmetric compounds is a prerogative of life that manifests itself in ordered organisms, in which the symmetry shapes the general layout and cells multiplication [9]. The last one allows the genetic heritage with the duplication by cells' symmetrical division. The asymmetry of forces determines lines of weakest resistance along which the growth is greater. D'Arcy W. Thompson explains the morphological variety by obstacles to the action of growth, which by nature tend to be uniform and symmetrical. He demonstrates that the growth process exhibit asymmetries in the balance of internal and external forces, which determines lines of weaker resistance that are critical in the diversification of forms. The balance of forces alters the speed of growth of different parts of the same body. In the small dimension of cells they are subject to the balance between internal and external pressure, then to the law of the surface tension on the cell's membrane; in the great dimension of vertebrates they obey to gravity and to mechanic's laws.

Sometimes the growth maintains the beginning form, such as in snails. More often asymmetries and alterations of the growth speed produce variations that lead to the diversification of final structures.

The morphologists adopt deformations of a Cartesian grid to describe the transformation among similar individuals. Those homological grids 'measure' the differences between two corresponding points in different species. Variants and invariants among species can be explained with diagrams that show formal homologies among individuals. Sets of key points that are arranged in lines of variable bending, correspond to these topological transformations [10].

4 The Organic Reference in Design

Arts and Design applied the reference to nature's models in different ways. Since antiquity man was fascinated by the beauty of the natural world, regular conformations of crystals, simplest living creatures, up to the Renaissance, when the human body was the expression of natural perfection. The first approach was the primitive imitation in the pattern of ornaments and decoration, due to the admiration of the beauty of natural harmony of regular conformation in rocks, plants and animals. Because of the formal identity of fundamental entities a close relationship links geometry and architecture. It is evident in the articulation of building, in which different applications of mathematical models resolve problems of both statics and aesthetics.

Later architects found in nature the solution to specific problems.

The architectural Orders are a fundamental example, as well as the most famous, but they are just one in between many. Indeed nature offers a great quantity of references to structural, formal and aesthetic problems. The perfect layout of many natural conformations led to the obsessive search for a rule of beauty based on geometry, meaning number and shape. Along centuries, the imitation of nature's objects is not a trivial game without any rational reason, as it seems, but the repetitive reference to nature's perfection expresses a rule for aesthetic equilibrium. Therefore it became the best design model, and it still is, because of reasons that natural sciences may explain (Fig. 4).

Since the principal motive for these similarities between structures depends on the force of gravity—which subjects all bodies both natural and built to the same laws of equilibrium—it is not surprising to find similar static schemes verified by analogous mathematical models. The model of the structural system can be as static (shells and ribs) as it is mechanical (the skeletons of vertebrates) and sometimes the natural architecture is much more complex than the manmade architecture, since buildings require neither movement, nor velocity.

Architecture applied the model of skeletons to structures, imitating nature's balance in the proportioned relationship of building elements. Later it was the connection of parts in machines. Finally the imitation is fulfilled in the process of growth, which is the expression of life. In its digital procedures, the responsive Design copies vital process of life. The golden ratio expressed a symbolic reference to natural growth, the procedural design goes further and it copies it, shaping artificial life.

Observation demonstrates that the relationship that exists between natural and artificial architectures, in the common composition of parts according to rules governed by geometry and/or the growth of forms, underlines the concrete nature of the classic myth of the imitation of nature. Because of the various symmetries that exist in natural forms, the effective foundation of this presupposition is indeed geometry, capable of conferring harmony and equilibrium. It therefore becomes an important element of design and construction.

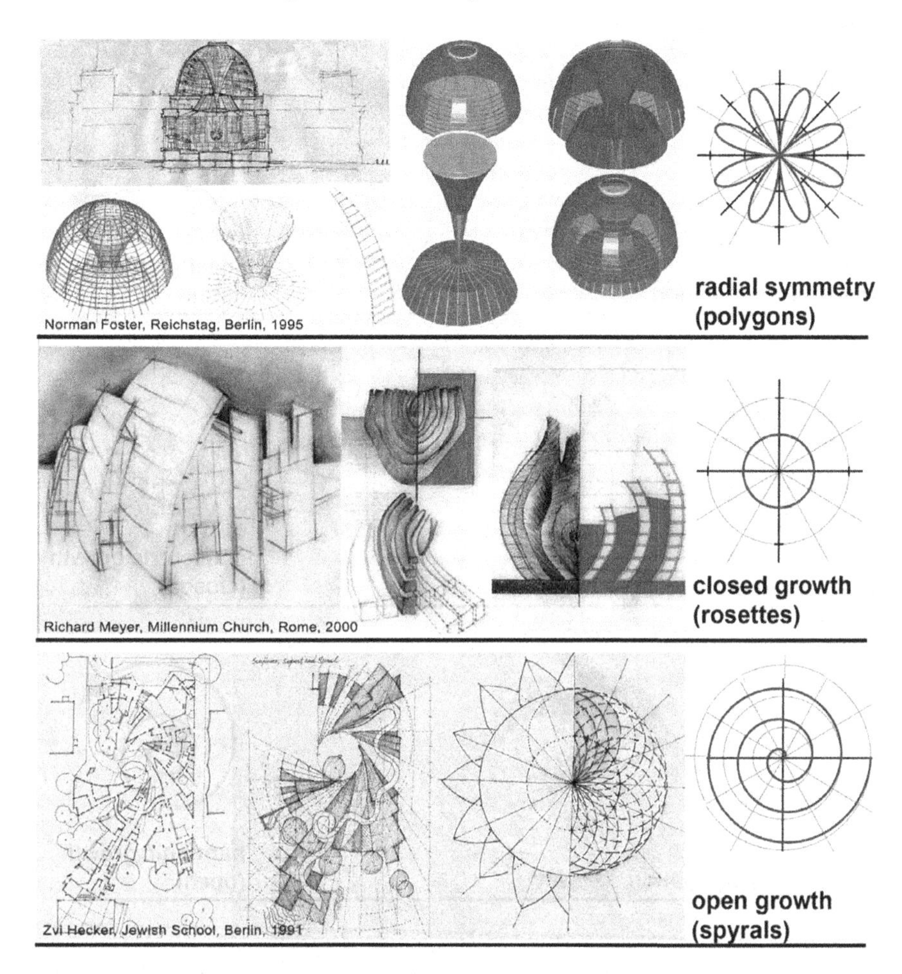

Fig. 4 Central pattern in architecture's forms with organic reference (student drawings). Circular patterns generate closed growth by increasing modules. Continuous growth becomes possible with spirals, which combine rotation and translation around a center

Good examples are M.C. Escher's and B. Fuller's work. The first "built the infinite" by the reiterate reproduction of cyclical divisions of the plan, which he applied eventually to the surface of the sphere using symmetry rules of five regular Plato's solids. His starting point was the cyclical tessellation of arabesque decoration, which recalls principles already defined in the late-nineteenth-century by ornament grammars. Those were also applied to wallpapers design. Escher extends the model to the pure concepts of surface and space, adapting plane patterns to the spherical continuous surface (Fig. 5).

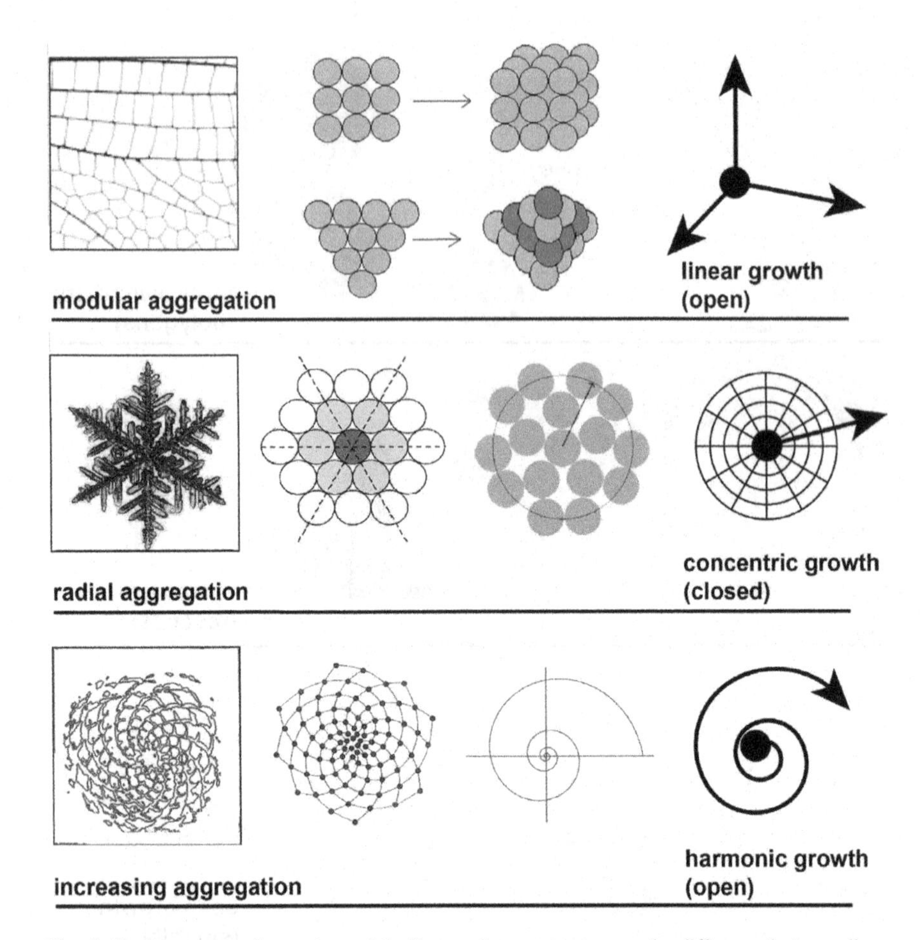

Fig. 5 Basic patterns of organic models. Patterns' structure stresses the difference between linear and radial aggregation of regular (uniform) cells: the first is open while the other is closed. Radial growth develops requires an increasing cells' number and produces net geometry only with hexagonal layout, otherwise it leads to aperiodical tessellations. (Penrose) Only logarithmic spirals allows regular (harmonic) growth

Escher's graphical inventions recall the works of Buckminster Fuller, who studied the spatial symmetries of the regular polyhedrons in order to solve structural problems. He applied regular pattern and lattices to the design of light structures, applying a mathematical program, deriving from the geodetic divisions of the sphere and/or the multiple symmetries of protozoa.

Effectively, mathematical models were developed to simulate reality by means of numbers, but geometry, which refers to form, is an concrete element of reality: everyone knows the logarithmic spiral of the Nautilus shell, the regularity of starfish, the perfection of the egg, and so on, but going beyond this, in protozoa are found living beings with the shape of all of the surfaces of Plateau, while radiolarian skeletons exhibit the forms of all five of the Platonic solids. It seems almost as though nature wanted to play with geometry.

Lars Spuybroek stated that computers have "*outgrown their servile function in the digital drawing room, where the real design was still done far away from the machines, sketched by hand, guided by genius*" [11].

The new computers' "cultural stage" gives new sap to the organic model in design, because they are able to go over the imitation of forms and structures. Computers master the process that lay behind the shape generation, such as the growth of living organisms or their adaptation to environment constrains. The imitation refers to life and its continuous transformation, not to the appearance of static forms. In nature "*all changes are small changes. Though a transformation can have a large effect, it is always a relatively small step, and the newness of the new can never be appreciated right away*" [12].

Any transformation applies the language of patterning, which bases on the four simple principles of symmetry. It can reach a higher level of complexity since bodies have internal and external transformation that interact with effects like merging, hooking, crossing, sliding, opening, nesting… The great variety of configurations in nature can be correlated to relatively few formal models based on different diagrams and symmetries, which make up the geometric basis of architectonic imitation. Both plane and spatial figures are always organised according to a simple diagram that can be traced back to three fundamental archetypes (Fig. 6):

- Modular aggregation according to a regular grid;
- Radial division of a circular unit in polygons;
- Linear continuity of spirals as regular growth of forms.

These models exhibit different kinds of symmetry and logic in particular growth patterns, and each of them has specific geometrical rules. Modular aggregation permits growth that is discontinuous and asymmetrical, according to the direction and the number of grid lines; it recalls histological tissue and can cover the plane and fill space, as well as expand linearly.

The symmetries according to which the base module is reproduced are four (translation, reflection, rotation and rototranslation). They combine with each other in very complicated ways, but they are always repetitive. The module is predetermined in relation to the fundamental grid diagram, but does not preclude a great variety of solutions (Fig. 7).

Growth is conditioned by the module, and this modifies complex form of the whole, which is indeterminate and thus permits the greatest degree of liberty. We can observe these models in the drawings of surfaces, in relationship to ornament and wall structure, and in the modular aggregation of spaces in plan as well as in spatial composition. We find them in the shapes of surfaces, in the structural mechanics of constructions, and again as an ambiguous game between the drawing of the surface and the representation of space. Radial division exhibits a closed form and a repetitive symmetry with respect to the centre, which often has mirror symmetry, but not necessarily the same number of axes. Growth takes place only in an outward direction, thus it is discontinuous and remains concentric so that the

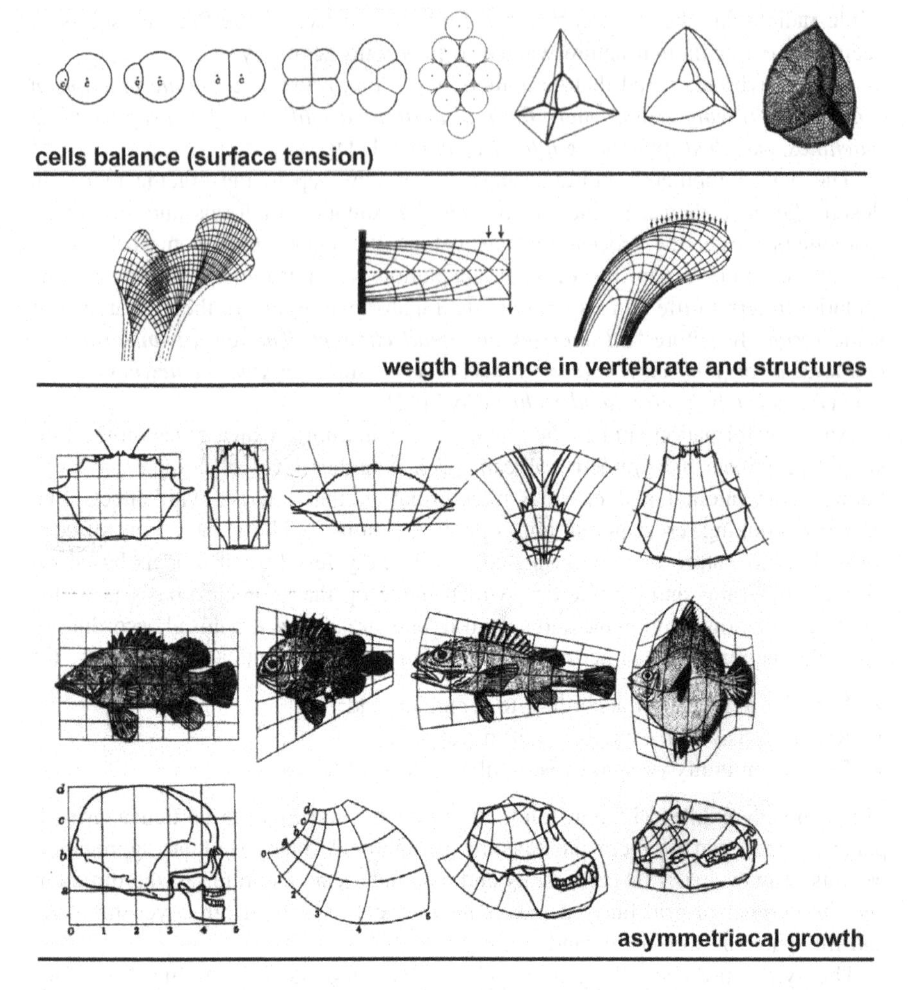

Fig. 6 Shapes and forces. Nature's conformations show as the (regular) shape follows the balance of forces: pressure on cells surface (surface tension), weight and gravity in vertebrates and structures, differential growing in biological variety (D'Arcy W. Thompson)

form is predetermined. This model can have a radial grid or can be aggregated in relationship to other grids. The extensions in space of this model are identified with rotated closed solids, such as domes.

Spirals derive from unidirectional linear growth that can be either planar or spatial, but which is usually refers to a continuity that tends to the infinite and to a particular rotational symmetry, which in logarithmic curves does not alter the proportions of the form. Thus growth is not discontinuous, and the form remains open.

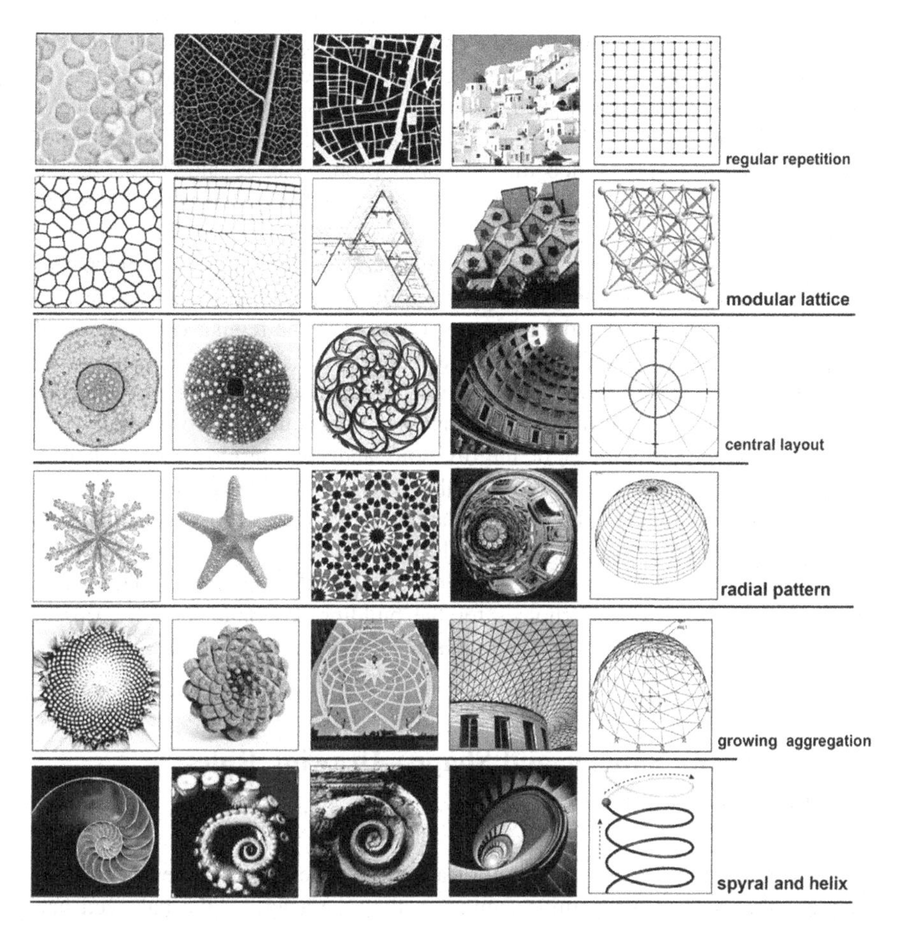

Fig. 7 Natural models and Geometry's rules. Basic conformations depend on cells and molecules aggregation in tissues and crystals, following basic patterns that develop different aggregation and growth possibilities. Design solutions refer to and interpretate organic forms in ornament, shape and layout

As a consequence of these conditions, man seems to have been particularly inspired by spirals since antiquity: spirals are used as special symbols of life and are manifest in art in different ways, while in architecture they become actual architectural elements.

Their spatial conformation generates complex surfaces that derive, however, from the regular motion of simple forms. Thus, since they are based on geometric rules, it becomes easy to draw, understand and construct the form.

These three models can be combined with each other in infinite combinations, each of which can be in its turn varied while maintaining homologous characteristics. These geometrical schemes satisfy various requisites of construction, suggesting solutions both formal and structural, with numerous examples found at all scales in architecture, from urban and territorial planning to ornament and surface decoration. We all know the significance that the concept of the module has in architecture and its importance in measuring, which is precisely the relationship between unit and quantity. Since this concept is directly connected to the use of modular grids that govern composition and proportion, it can be said that the design project makes reference to the concept of measure, and that this takes place through geometry.

The module is the basis of architectural order, which is the first principle of structures, organised according to spatial grids with orthogonal directions. In architecture cubic and pyramidal grids are common; in the regular organisation of the plane as well as the articulation of surface there are many possible solutions. In spite of its being a closed form, solutions based on the radial scheme are numerous and diverse; in drawings of plane configurations, such as those of rose windows or pavement designs, the number of the divisions and concentric elements change. This plane scheme is often used in urban design and in the realisation of buildings with a centralised and hierarchical spatial layout, in which the formal articulation is reflected in adjunct minor spaces. In spatial forms this scheme generates domes that can be composed of surfaces conceived according to different design solutions, in relationship to the structural choices for the building: continuous shells, ribs or geodesic grids as in the work of Buckminster Fuller.

5 Conclusion: Number and Shape—Drawing Rules and Design Codes

A regular lattice rules the aggregation of atoms in molecules and then in of more complex structures in a repetitive symmetry that characterizes the balanced harmony of natural forms. The ordered beauty of geometrical configurations of nature that caused wonder and admiration offered the main model to Man's creativity and first inspired decoration and artificial building. These ancestral models are attributable to the divisions of the unit in regular polygons that are inscribed in circles, to the continuous growth of the spiral or in the additions of gnomons and to the juxtaposition of repetitive modular elements in lattices. The basic rules of form are the same simplest operations of arithmetic and geometry: *division*, *multiplication* and *addition*. In the classical world the formal beauty was linked to a recognizable law that order the multiplicity in the unity: *symmetry*, *proportion* and *direction* resume rules that generate the shape. Together they express the *eurythmy*, which Gottfried Semper defines as a "*concatenated sequence of spatial ranges, similarly shaped*" [12].

Later the imitation recalled the relationships among elements in a closed system, just as Buckminster Fuller and Frei Otto did. Contemporary designers still explicitly refer to nature and its transforming processes, according to Lars Spuybroek's statement. The procedural design by generative software allows the digital imitation of organic models is not only a formal reference such as in its historical roots. It just pursues the dynamic transformation of shapes in living process. Generation of shape starts from a beginning tile and a developing concept. The form comes from the repetition of reiterate algorithms that change the shape adding new elements in the tessellation, turning the design into a process of biological growth. The configuration is the goal of a dynamic system that is regulated by predetermined codes, which are able to adapt themselves to boundary conditions, interacting with the surrounding environment. Similar shapes result from the management of a small number of formal parameters.

The dynamic model provides a even complex organization in which the rules of design meet to the homeomorphisms of the topological geometry.

Symmetry implies stability and balance; however it contrasts with the tension of growth and changing of living organisms that appears in the asymmetry of topological transformation. The imitation of Nature accomplishes in the continuous growth of generative design.

6 Notes

1. Ernst Haeckel, 1868.
2. B. Russel, 1959.
3. Plato, Fedon.
4. D'Arcy W. Thompson, pp. 126–129 (Italian transl.).
5. Plateau ha demonstrated that exist only 6 minimal surfaces that have 1 symmetry axe: sphere, plane, cylinder, catenoid, nodoid and onduloid. D'Arcy W. Thompson, 1917.
6. Keplero, Harmonices Mundi, 1619.
7. Plateau solved the Lagrange's problem (minimal surface depending on perimeter or volume) with sheets of soap water. In protozoa exist the 6 Plateau's surfaces. See D'Arcy W. Thompson, cit., cap. 3.
8. D'Arcy W. Thompson, cit., cap. 5.
9. Pasteur noted that the asymmetric structure is one of the deeper aspects of the difference between vital phenomena and not in life: "*this is perhaps the only line of demarcation that marks the difference between the chemistry of living matter and of matter that is not alive*". See D'Arcy W. Thompson, who stresses that in Nature the life develops often from a simple tube.
10. The Osaka Group states that the Nature be inclined to predetermined patterns. See J. Fodor, M. Piattelli-Palmarini, 2011.
11. L. Spuybroek (2004), p. 9.
12. L. Spuybroek (2004), p. 4.
13. Gottfried Semper, 1860. See the introduction.

References

1. Semper, G. (1860–1863). *Der Stil in den technischen und tektonischen Künsten*. München.
2. Haeckel, E. (1868). *Natürliche Schöpfungsgeschichte*. English trasl: The History of Creation, 1914.
3. Thompson D'arcy, W. (1917). *On growth and form*. Cambridge University Press. Italian trasl: Universale Bollati Boringhieri, 1992.
4. Russel, B. (1959). *Wisdom of the west: A historical survey of western philosophy,* Italian trasl: Longanesi, Milano, 1978.
5. Fuller, R. B. (1975). *Synergetics. Explotations in the geometry of thinking*. New York: Macmillan Publishing.
6. Stewart, I. (2001). *What shape is a Snowflake?* London: Weidenfeld & Nicholson. Trad. it. Che forma ha un fiocco di neve, Bollati Boringhiari, Milano.
7. Spuybroek, L. (2004). *Nox: Machining architecture*. Thames & Hudson, Londra.
8. Rossi, M. (2006). Natural Architecture and constructed forms: Structure and surfaces from idea to drawing. *Nexus Network Journal, 8*(1), 112–122.
9. Nature Design. (2007). *From inspiration to innovation*. Museum für Gestaltung, Zurigo, Lars Müller Publishers.
10. Ferrero, G., Cotti, C., Rossi, M., & Tedeschi, C. (2009). Geometries of the imaginary space: Architectural developments of ideas of M.C. Escher and Buckminster Fuller. *Nexus Network Journal, 11*(2), 305–316.
11. Hays, K. M., & Miller, D. (Eds.). (2008). *Buckminster Fuller, Starting with the universe*. New York: Whitney Museum of American Art.
12. Fodor, J., & Piattelli-Palmarini, M. (2011). *What Darwin got wrong*. Profile Books.

Shapes of Design: Traditional Geometry, Symmetry and Representation

Giuseppe Amoruso

Abstract Geometry, traditional arts and design universally contribute both to the formation of ideas that to the creation of form by establishing a cultural contract between art and science. The introduction of digital design media has been imposing new and practical ways to rediscover knowledge or culture in representation and design from traditional arts. The subject of geometric tiling of the Euclidean plane is always a contemporary challenge by combining creative roots of artistic and scientific research. From the ornamental motifs of the Alhambra in Granada and the Mezquita of Cordoba, geometry, shape and colour allow designers to experience a wide range of meanings, ranging from the practical and technical sphere, to the spiritual and sacred one.

Keywords Traditional arts · Geometry · Symmetry · Design pattern · CAD

1 Introduction

Science and art form the cultural foundations of a designer, because they share common roots, the first addresses the public experience, universal, objective, quantitative, unitary, and its language is precise, rational, made of ideas and concepts. The second looks at the experience rather private, particular, subjective, qualitative, and its language is ambiguous, emotional, made of images and stories [1]. The connection between the two cultural spheres is in mathematics and geometry which generate languages and representations (Fig. 1).

Among the Greek philosophers, Plato, illustrates his aesthetic thought that was the language of mathematics. The Philebus is a dialogue written in the final stage of its production (366-365 BC), in which the philosopher attributes to the master the leading role: discussing with Philebus and Protarco, Socrates seeks the "*true Good*"

G. Amoruso (✉)
School of Design, Politecnico di Milano, Milan, Italy
e-mail: giuseppe.amoruso@polimi.it

M. Rossi and G. Buratti (eds.), *Computational Morphologies*,
https://doi.org/10.1007/978-3-319-60919-5_3

Fig. 1 Geometric construction starting with the star-and-hexagon design

that can guarantee a happy life, starting with the possibility—later denied—that it coincides with the pleasure.

In the dialogues between the philosophers we explore the concerns of beauty of form that is matched to the shape of finite, those geometric ones achieved by the same antique tools, a compass, a straight edge and a T-square. These dissertations open the way to the concepts of proportion, harmony, beauty and truth through memory and spiritual approach to life the art forms [2].

"*Art holds out the promise of inner wholeness*" wrote Alain de Botton [3] but perhaps the deepest argumentation on how arts serve the spirit came more than a century earlier, in 1910, when Russian painter and art theorist Wassily Kandinsky published Concerning the Spiritual in Art—an exploration of the human and most original reasons for performing art, the "*internal necessity*" that move artists to create as a spiritual impulse and audiences to admire art as a spiritual hunger.

Kandinsky, who was greatly influenced by Goethe's theory of the emotional effect of colour and who was himself synesthetic, highlights the powerful psychic effect of colour, and its interactions, in the comprehensive spiritual experience of art [4].

Today the practice of the traditional arts is the starting point of generation of designers bringing their soul more close to the creative process based upon "*universal spiritual truths*" (Fig. 2).

The fundamental principle that has to be pinpointed is that tradition is a continual renewal when scholars, designers start their apprentice. The book of the universe which we learned about from Galileo is written in the language of mathematics and its characters are triangles, circles and other geometric figures. The Italian scientist was responsible for the birth of the modern research method, hypothetical and experimental that aims to formulate a scientific law, in the years when the ideas of Platonism had spread again throughout Europe and in Italy and

Fig. 2 Geometric pattern found in the Ibn Tulun Mosque in Cairo (879 C.E.)

probably also for this reason the symbols of mathematics are identified by him with geometric entities and not with numbers. Essentially, it was the contribution of Galileo to the scientific language, in particular, in the writings of Galileo many words are taken from the common language and undergo a process of assigning new and specific meaning (a form, then, of semantic neologism) [5].

Geometric design and the use of digital systems of representation demonstrate the validity of Galileo Galilei statement that nature is compared to a book; it is written in a language whose letters are polygons and circles, and using as pens the classical tools of Euclidean geometry, ruler and compass, following the universal tradition that goes from the early Greek geometers to the expressions of modern design.

The Platonic vision settled the fundamental research about "*ideas*", what we currently assume as shapes and forms; the unique way to develop meanings is to research the right geometry and, in this way, return to investigate about common roots, formulas and ratios coming from math.

This is the reason of the continuous success of the golden ratio in arts and design; the same common numerical and irrational code that it was conceptualized and built in form of a temple, a painting, a building type and a music instrument. A universal truth and knowledge that is often described as the essence of classical architecture where proportions have the meaning of equality of ratios and a number becomes "*divine*" according to Luca Pacioli [6].

Experiencing traditional arts as performing skills, designers participate in the regeneration of knowledge about the geometric universe (Fig. 3).

Fig. 3 Geometric rosette generation

The study aims to improve the awareness among young designers that form, pattern and colour as manifested in the various branches of the traditional arts and design, have a deeper meaning not simply pleasing to the senses, but playing a strategic role transferring both philosophical and technical contents. For example the relations between geometry and illusory design that was conceived and expressed in a multiple forms of representation: optical corrections, perspectival frescoes, perspective-relief and scenography (Fig. 4).

For centuries practitioners used always the same simple tools to design and build their concepts, from sacred architecture to dwellings and objects.

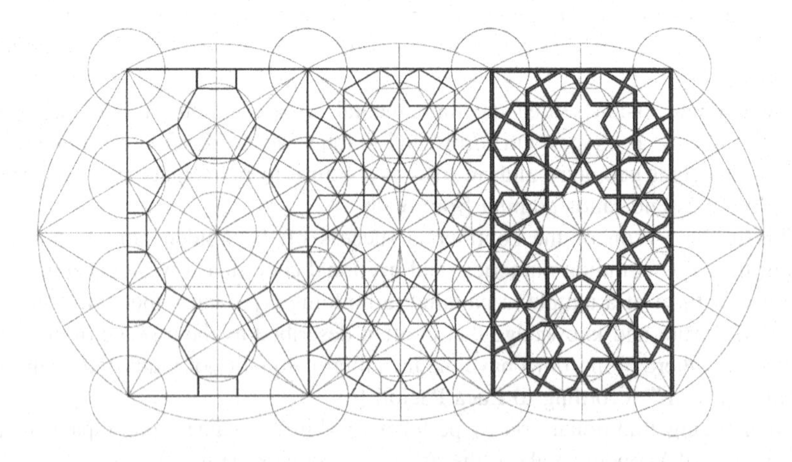

Fig. 4 Semiregular tiling of regular dodecagons, regular hexagons and squares

Whether a traditional art is representational (iconic) or non-representational (aniconic) it will always be based on the principles of ordered space. All civilisations have acknowledged that geometry is fundamental to the cosmic order, or as Plato explained, "*…geometry is the knowledge of the eternally existent*" (Republic, Book 7, 527b). The School teaches geometry, not only as an objective language informing the traditional arts of the world, but also as an essentially sacred language. Students learn that patterns of traditional art reflect nature and are underpinned by the same geometry that is the basis of the natural world. Thus geometry is seen as a reflection of a universal order, as was taught by the ancient Greeks and recognised by the great Arab architects and scientists, as well as the cathedral builders of the Middle Age.

Geometry intervenes in the creative and conceptual process in three forms: as a language, as a representation or as structure.

Symmetry, in the words of Hermann Weyl, both in the broadest sense and in specialized fields, it is an idea that has guided the man through the centuries to the understanding and the creation of order, beauty and perfection [7].

2 The Illusion of Infinity

Among the challenges that have always engaged mathematicians and creatives is the regular division of space that has found a special contribution from Escher that studied and drawn formal and aesthetic solutions influenced by his discoveries of Islamic patterns in Granada, Sevilla and Cordoba. It is the theory of tessellations, a graphic process that tiles a plane in a finite number of small elements obeying to geometric rules related to the concept of symmetry. This idea has inspired the research of meanings and forms combined with multiple functions, involving artists, scientists and philosophers. This is the universal necessity of representing balance and harmony through the language of the form that includes a deep mathematical theory that describes a phenomenon. First templates of tessellations were found in some Sumerians buildings (about 4000 BC) as primitive geometric wall decorations built as patterns of clay tiles. The first documented studies about tessellations are referenced to Johannes Kepler when, in 1619, he published his Harmonices Mundi (Harmony of the Worlds); he presented regular and semiregular tessellation to be assumed as covering of a planar surface with regular polygons. In 1891, the Russian crystallographer Yevgraf Fyodorov proved that every periodic tiling of the plane features one of seventeen different groups of isometries, surprisingly used by Egiptians in the Valley of the Kings and by Arabs in the Alhambra's decoration. Fyodorov's work, in the field of geology, marked the beginning of the mathematical study of tessellations. Other prominent contributors include Shubnikov and Belov (1951); and Heinrich Heesch and Otto Kienzle (1963), but also Rosen (1995) that studied the modern mathematical theory of symmetry and its deep influence in science branche (Figs. 5, 6).

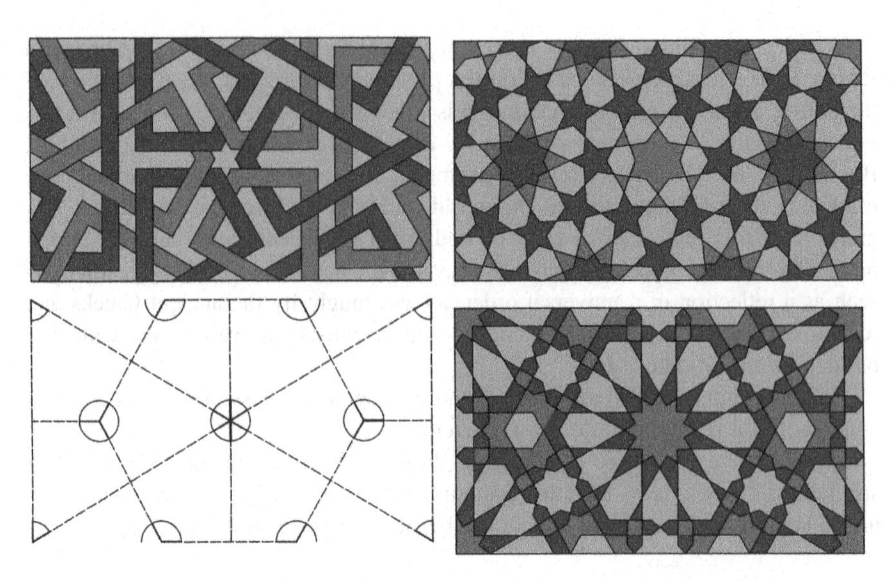

Fig. 5 Three-fold permutations: hexagonal grid (or equilateral triangle) generated patterns

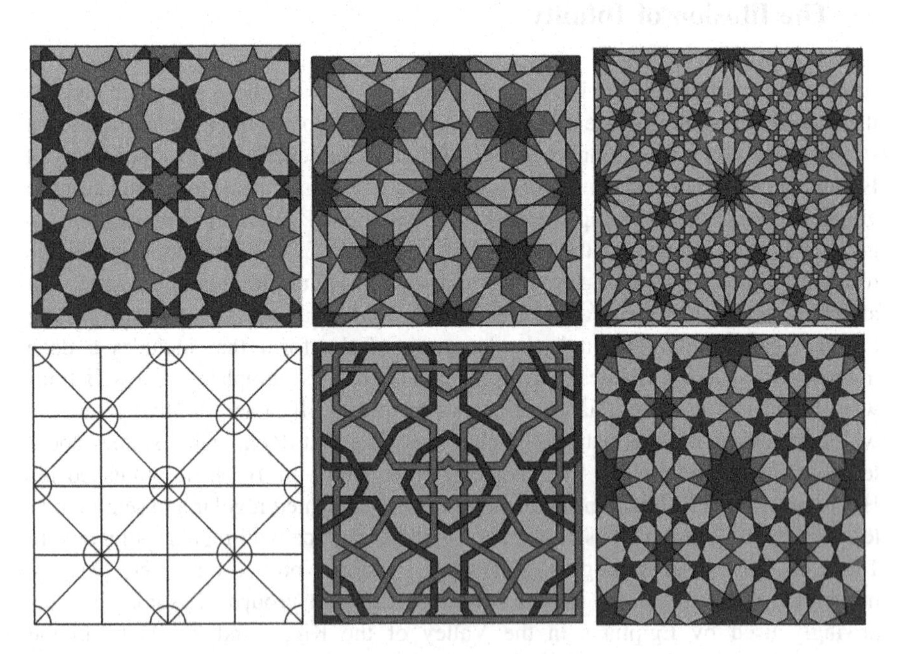

Fig. 6 Four-fold permutations: square grid generated patterns

Unfolding geometrical elements from the starting concept of unity, it starts the creative process and the repetition of patterns where the loop can continue indefinitely; Sutton describes this solution as "*the perfect visual solution to calling to mind the idea of infinity, and hence the Infinite, without any pretence of being able to truly capture such an enigmatic concept visually*".

Transforming shapes into psychological and emotional effects reminds us the concept of optical illusions, well known since the early designers steps and that have appealed to the mind of spectators throughout history, and have had great impact when combined with real architectural elements or spaces. Illusionary methods in design have been used by artists and architects since antiquity, but from Renaissance they were scientifically analysed with the invention of perspective and later in the baroque with the discover of anamorphosis, and their integration with arts: the architectural perspective known as quadratura and the illusory painted architecture of Andrea Pozzo in Rome.

Forms of tessellations and their use as artworks play, according to Spiliotis, the specific illusion of "*plane dematerialisation*" that Mediterranean and Eastern cultures have exhibited as "*patterned optical illusions*" on floors of tiles or woodwork which seem to deconstruct the flat floor plane and create additional illusory ones; templates have been found in a number of countries including Japan, China, India, Persia, and Italy.

The theoretical foundation is that of invariance under isometric transformations of the Euclidean plane (and space): named isometry when the bijective transformation does not alter the distance between two points. Isometries of the plane belong to four main classes, reflections, rotations, translations and glide reflections, establishing measurement tasks. Substituting figures with numbers we introduce algebraic operations defining "*composition laws*", as transactions between numbers, which allow to obtain the algebraic structure of a "*group*" and then of a "*group of symmetries*".

A periodic tiling produces a repeating pattern: some templates include regular tilings with regular polygonal tiles all of the same shape, and semi-regular tilings with regular tiles of more than one shape and with every corner identically arranged [8].

A plane symmetry group is a mathematical classification of a two-dimensional repetitive pattern, based on the symmetries in the pattern. Such patterns occur frequently in architecture and decorative art. The patterns formed by periodic tilings can be categorized into 17 wallpaper groups. Wallpaper groups are two-dimensional symmetry group, intermediate in complexity between the simpler frieze groups and the three-dimensional crystallographic groups (also called space groups).

Among the works of Escher we find himself curiously displaying the principle of psychological perception of symmetrical element that from the alteration of the geometric symmetry allows just to appreciate its value; this principle concerns a hierarchical sequence of elements or groups of symmetry found in natural forms or design solutions that allows to highlight their beauty or harmony.

According to Escher the rules of symmetry and harmonic balance are conceived, reproduced and broken building hidden meanings and introducing an additional degree of freedom: colour becomes the aesthetic key of altered symmetry [9].

Sometimes the colour of a tile is understood as part of the tiling, at other times arbitrary colours may be applied later. When discussing a tiling that is displayed in colours, to avoid ambiguity one needs to specify whether the colours are part of the tiling or just part of its illustration. This affects whether tiles with the same shape but different colours are considered identical, which in turn affects questions of symmetry. The four colour theorem states that for every tessellation of a normal Euclidean plane, with a set of four available colours, each tile can be coloured in one colour such that no tiles of equal colour meet at a curve of positive length. The colouring guaranteed by the four-colour theorem will not in general respect the symmetries of the tessellation. To produce a colouring which does, it is necessary to treat the colours as part of the tessellation.

Finite groups of symmetry belong to the geometry of rosettes, templates of radial symmetry, special works of art embodying issues of symbolism and fascinating beauty; in the rosettes geometry all rotations must have the same centre according to a principle that Weyl attributed to Leonardo who also dealt of symmetry in architecture.

However one of the moments of maximum splendour of radial symmetry belongs to the Islamic art practice [10]. The prohibition of representing holy figures led the figurative Islamic culture to move towards geometric patterns [11]. From Middle Ages another artistic representation of radial symmetry can be seen in the cosmatesque floors, a style of geometric decorative inlay stonework typical of the architecture of Italian Medieval churches and derived from the Byzantine Empire templates dealing with the geometry of circles and squares.

This geometric tradition finds multiple expressions in the design of "*mandala*", (in Sanskrit maṇnala means circle), that is a spiritual and ritual symbol in Hinduism and Buddhism, representing the Universe but also the best image to visually frame the infinite into the finite.

A variant of the Mandala are the "*kolam*", drawings made at the entrance of the house by Indian women with auspicious function.

Both mandala that kolam can be drawn starting from intersections of circles and squares and then stained freely. It is an exercise very suitable to approach both geometry and harmony of shapes and colours which is free from figurative representation [12].

3 Conclusion

The study of traditional arts and geometric principles that from time to time assign functional, symbolic, philosophical and spiritual meanings still represents a living challenge.

Hidden geometries and their degrees of complexity intervene in the structure, in the language and in the representation of design shapes, guiding the mature and advanced creative process.

The research about symmetry holds a special place in the generative processes that now constitute a new challenge for designers when digital systems of representation are introduced.

This contribution is a simple step towards a return to the study of geometry as a multimedia environment that today the use of digital technologies in the design and also generative software can renovate delivering innovative forms of mind.

Acknowledgements Illustrations from Giuditta Margherita Maria Ansaloni (Fig. 1), Elena Cattani (Fig. 2), Marco Dassi (Fig. 3), Federica Cocco (Fig. 4), Silvia De Bellis (Figs. 5, 6) according to the studies and methodologies presented by Daud Sutton in Islamic Design. A genius for geometry, Wooden Books, Glastonbury, 2007.

References

1. Odifreddi, P. (2006). *Penna, pennello e bacchetta. Le tre invidie del matematico*. Bari: Laterza.
2. Zadro, A. (1982). *Filebo*. In *Platone Opere complete* (Vol. 3). Bari: G. Giannantoni, Laterza.
3. De Botton, A., & Armstrong, J. (2013). *The school of life, art as therapy*. London: Phaidon Press.
4. Kandinsky, W. (1946). *On the spiritual in art*. New York: Guggenheim Foundation.
5. Galilei, G. (1623). *Il Saggiatore*. Rome: Accademia dei Lincei.
6. Pacioli, L. (2010). *De Divina Proportione. Riproduzione anastatica della copia conservata presso la Biblioteca Ambrosiana di Milano*. Milano: Silvana Editore.
7. Weyl, H. (1952). *Symmetry*. Princeton: Princeton University Press.
8. Wade, D. (2006). *Symmetry. The ordering principle*. Glastonbury: Wooden Books.
9. Grasselli, L., & Costa A. (2014). Le forme della simmetria: dai mosaici dell'Alhambra ai mondi di Escher. In M. C. Escher, & M. Bussagli (Eds.), Skira, Ginevra-Milano.
10. Sutton, D. (2007). *Islamic design. A genius for geometry*. Glastonbury: Wooden Books.
11. Azzam, K. (2013). *Arts & crafts of the Islamic lands: Principles materials practice*. London: The Prince's School of Traditional Arts, Thames & Hudson.
12. Cunningham, L. B. (2010). *The Mandala book: Patterns of the universe*. New York: Sterling.

Pattern Spaces: A Rule-Based Approach to Architectural Design

Marco Hemmerling

Abstract The paper discusses the potential of patterns for the architectural design process, both as an aesthetic reference as well as a design methodology. The latter one, as a way of describing and communicating design, is highly associated with computational methods in architecture, such as generative and parametric design. Against this background two case studies, as part of an academic project, exemplify a rule-based design strategy—from pattern analysis to programmable three-dimensional structures. This bottom-up approach allows not only for the generation of complex, yet controllable spatial solutions, but enhances also the design decision process through the creation of manifold variations.

Keywords Generative design · Rule-based design · Pattern · Design methodology

1 Introduction

Patterns have been an integral part of Architecture since ancient times. They form the basis of the history of ornament, an aesthetic phenomenon that links all times and cultures at a fundamental level. Obviously styles of ornamentation are connected to their cultural background and therefore carry different functions, meanings and aesthetic expressions. Yet, all these patterns are based on geometric principles. They consist of, or are generated from, simple forms like a circle or a square. Through algorithmic operations—such as rotate, copy, array, mirror, offset and scale—intricate and fascinating patterns have been developed from these basic geometries. Patterns have an underlying mathematical structure; indeed, mathematics can be seen as the search for regularities, and the output of any function is a mathematical pattern. The architecture avant-garde in the first half of the twentieth century as well abolished ornament and patterns from architecture. The infamous manifesto "Ornament and Crime" [1] by Austrian architect Adolf Loos influenced

M. Hemmerling (✉)
Faculty of Architecture, Cologne University of Applied Sciences, Cologne, Germany
e-mail: marco.hemmerlinf@th-koeln.de

M. Rossi and G. Buratti (eds.), *Computational Morphologies*,
https://doi.org/10.1007/978-3-319-60919-5_4

the minimal massing of modern architecture, and stirred controversy. He explored the idea that the progress of culture is associated with the deletion of ornament from everyday objects. Nevertheless, the notion of patterns has taken on new meaning and importance since the 1960s. US-American architect Frank Lloyd Wright defined ornament as an integral part of architecture when he stated:

> True Ornament is not a matter of prettifying externals. It is organic with the structure it adorns, whether a person, a building or a park. At its best it is an emphasis of structure, a realization in graceful terms of the nature of that which is ornamented.

Christopher Alexander brought in a new perspective to patterns through his book "A pattern language" [2]. In this influential treatise he described 253 design patterns that address specific architectural problems and offer at the same time practical design solutions. Most notably, none of the patterns is an isolated entity. All of them are interconnected and form a holistic network with manifold connections. Alexander defines architecture as a system, which has often been associated with information technology. Hence, the idea of a pattern language appears to apply to any complex engineering task, and has been applied to some of them. It has been especially influential in software engineering where patterns have been used to document collective knowledge in the field. Christopher Alexander himself described the complex behavioral relation of patterns as follows:

> No pattern is an isolated entity. Each pattern can exist in the world only to the extent that is supported by other patterns: the larger patterns in which it is embedded, the patterns of the same size that surround it, and the smaller patterns which are embedded in it.

Complexity research has ultimately shown that even highly complex, dynamic patterns may be based on simple behavioral rules, and that has allowed the notions of pattern and pattern formation to take on new meanings, that are also central for architecture. Today the use of generative computerized methods is opening up new ways of talking about an idea that is becoming increasingly abstract and dynamic. Patterns allow for the efficient tessellation of a surface or volume, based on a minimal number of elements. Furthermore, computational tools enable us to generate complex geometries while maintaining control over both the tessellation system (pattern) and the overall form. The interplay between system and form creates an interesting problem for architectural design: System follows form or form follows system?

2 Case Study: Pattern Spaces

Against this background the academic project "Pattern Spaces" aimed at a rule-based approach to architectural design that addresses both—the aesthetic and functional qualities of a pattern as well as the idea of a behavioral principle as a design methodology. The project was organized in a bottom-up process, starting with the geometric, functional and aesthetic analysis of a self-selected pattern. The re-combination, extension and interpretation of the elements was then carried out first in 2D and later in 3D, based on the findings of the analysis. At this point the set-up of a flexible design environment through visual programming (Rhinoceros/ Grasshopper) was an essential step for the further development. During this phase the focus was clearly on the design of the process, rather than on the form-finding. Hence, the exploration of variations throughout this generative process proofed to be an essential instrument of design decision-making. As a final step the developed spatial structures were interpreted as an architectural design, taking into account functional, aesthetic and structural qualities as well as aspects of materialization, usage and perception. The following two case studies carried out by Edgar Hildebrand (Penrose Tiling) and Kristin Osthues (Axonometric Projection) are representative for 50 individual students works as part of the project "Pattern Spaces".

2.1 Penrose Tiling

The project by Edgar Hildebrand is based on the Penrose tiling introduced by mathematician Roger Penrose in the 1970s. It is an aperiodic pattern, which implies that a shifted copy of the pattern will never match the original. It may be constructed so as to exhibit both reflection symmetry (mirror) and fivefold rotational symmetry (rotate), but it lacks any translational symmetry (move). The pattern is self-similar like a fractal geometry, so the same patterns occur at larger scales. Furthermore the Penrose tiling is a quasi-crystal and can continuously fill all available space (Fig. 1).

The tiles of a Penrose pattern are constructed from shapes related to the pentagon, but the basic tile shapes need to be supplemented by matching rules in order to tile aperiodically. The given pattern uses a pair of rhombuses with equal sides but different angles. Ordinary rhombus-shaped tiles can be used to tile the plane

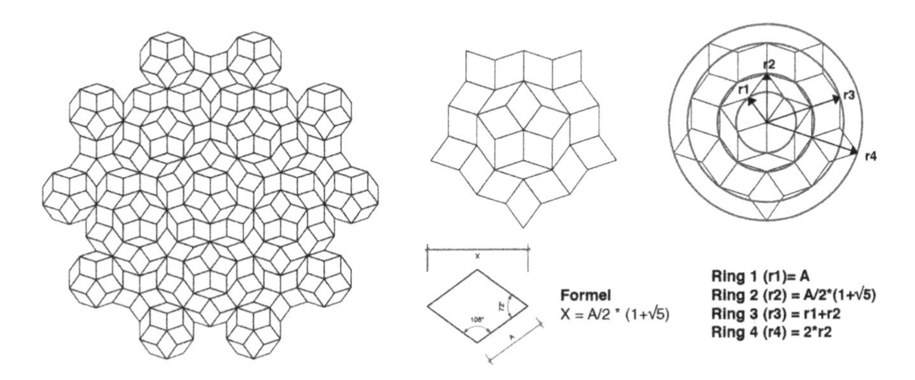

Fig. 1 Penrose tiling, geometric principles

periodically, so restrictions must be made on how tiles can be assembled: no two tiles may form a parallelogram. The geometric relation between the sides of the rhombuses can mathematically be described by the equation: $x = \mathrm{A}/2 \times (1 + \sqrt{5})$. In order to develop a computational model, based on visual programming a geometric description of the Penrose systems had to be defined. As a result of the analysis the interdependency of the pattern was found in concentric rings that allowed for the definition of the various rhombus vertices as shown in Fig. 1 on the right. Based on these constraints the programming allowed for a manipulation of the vertices in X- and Y-direction by scaling the diameter of the particular ring and in Z-direction by defining a height value for each individual point (Fig. 2).

During the design process many variations were carried out to test possible spatial solutions (Fig. 3). From these variations the tectonic idea for an outdoor pavilion was developed further, taking into account the scale and position of the spatial design. In addition a factor to define openings in-between the resulting double-curved surfaces was introduced to control the light ambiance under the

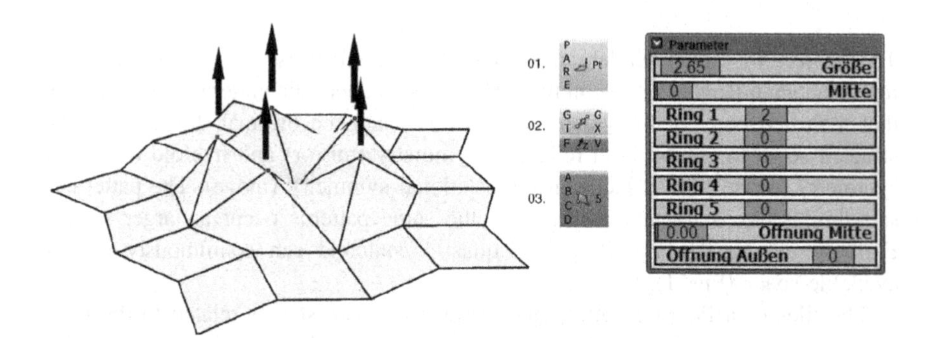

Fig. 2 Penrose tiling, computational model

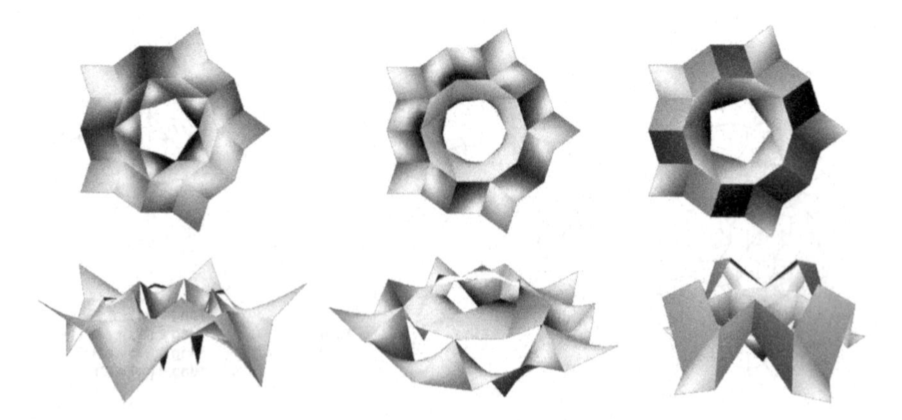

Fig. 3 Penrose tiling, spatial variations

pavilion structure. In a final step two concepts for construction and materialization were investigated—a membrane structure—based on textiles and tensile components, as well as a triangulated structure—made from flat wooden panels.

2.2 *Axonometric Projection*

Axonometric patterns are often to be found in paving tiles. On one hand they allow for an efficient surface layout and on the other hand they create an interesting three-dimensional illusion on a two-dimensional plane. Accordingly, Kirstin Osthues' interest was driven by the effect that axonometric patterns have on the viewer and the interplay between 2D abstraction and 3D perception. Axonometric projection is a type of parallel projection where the plane or axis of the object depicted is not parallel to the projection plane, such that multiple sides of an object are visible in the same view.

Depending on the exact angle at which the view deviates from the orthogonal, three basic types of projection can be distinguished: isometric, dimetric and trimetric projection. All types of axonometric projection display the parallel edges of an object as well parallel (Fig. 4)—unlike the perspective projections, which are constructed from vanishing points.

The first part of the project was dedicated to the analysis of the underlying geometric principles that constitute different axonometric projections. In a second step common features and strategies of transformation from one to the other were investigated As a result the findings were translated into a computational model that allowed for a three-dimensional transformation of each type of projection while keeping at least one view appearing as the original pattern (Fig. 5). In addition the programming of the geometry enabled a gradual transformation between the different axonometric projection patterns.

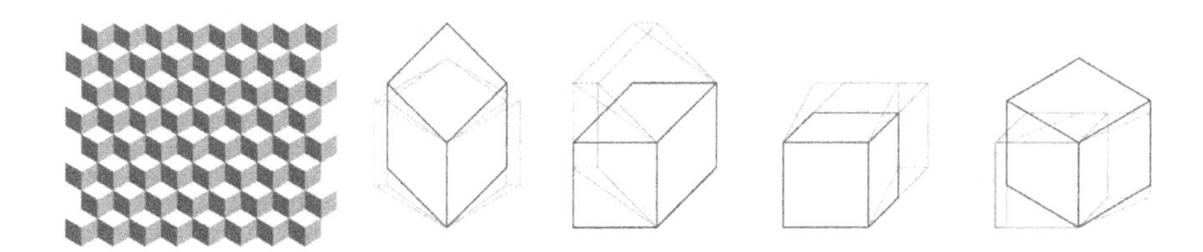

Fig. 4 Isometric tile pattern (*left*), axonometric projection types (*right*)

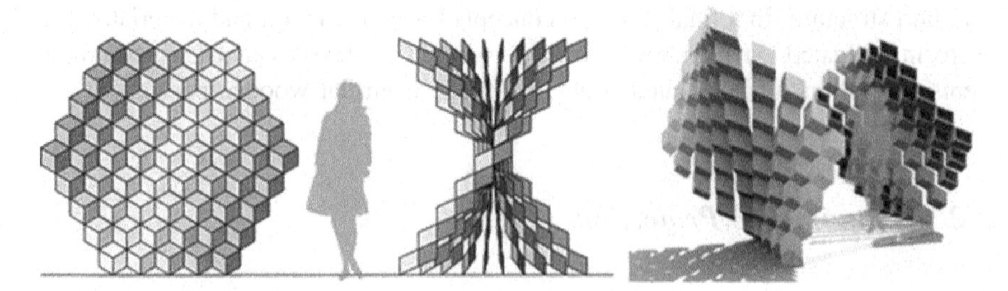

Fig. 5 Spatial interpretation of the computational model

3 Conclusion and Outlook

While the geometric complexity of the projects increased throughout the design process (from 2D to 3D), the students became not only proficient in the use of computational methods and tools, but foremost developed an in-depth understanding of the patterns' underlying principles and interdependencies. Formulating design rules and constraints enhanced abstract thinking and broached the issue of "how we design" rather than "what we design". In other words: by focusing more on the design of the process than on a formal result, each project generated multiple outcomes, which are interesting both in their disposition and phenomenology. Hence, they address, as mentioned before, the interplay of system and form. The two case-studies presented in the paper will be continued towards realization within the framework of the two research platforms ConstructionLab and PercpetionLab at the University of Applied Sciences Ostwestfalen-Lippe. While the ConstructionLab focuses on building technologies and construction methods, the PerceptionLab investigates the impact of space on the human being.

References

1. Loos, A. (1910/13). *Ornament and crime*, Cahiers d'aujourd'hui No. 5.
2. Alexander, C. (1977). *A pattern language*. New York.
4. Pottmann, H., a.o. (2009). *Architectural geometry*. Vienna.
5. ARCH + 189. (2008). Entwurfsmuster, Aachen.
6. ConstructionLab, http://www.hs-owl.de/fb1/en/forschung/construction-lab.html.
7. PerceptionLab, http://www.hs-owl.de/fb1/en/forschung/perceptionlab.html.
8. Rhinoceros/Grasshopper, www.rhino3d.com, www.grasshopper3d.com.

Algorithmic Modelling of Triply Periodic Minimal Surface

Giorgio Buratti

Abstract In the planning process design has always preceded the construction phase. The act of designing is an opportunity to organise one's ideas, manage resources and predict results, and is made possible through the use of dedicated instruments. Pencils, pens, compasses and other simple instruments have slowly been refined, remaining largely unchanged over the centuries, and it is only in the last four decades that they have been supplemented by computer systems. Over the years the level of involvement of computer grew, developing from a representational role to having a direct influence on the process of generating forms. In recent years, the increased levels of computer literacy has given rise to a new type of modelling, based on the elaborative logic of information, which has determined a new phase in computer assisted design, in which the form is generated by drawing up algorithms. In this paper this kind of modelling is applicate to the study of minimal surfaces. These geometric objects, if structured in a triple symmetry periodic system, reveal highly interesting properties that are hard to analyse and manage with traditional tools.

Keywords Design · Generative · Minimal surfaces · Algorithmic · Computational

1 Introduction

In recent years, many scientific disciplines have been turning with great interest to the study of minimal surfaces. This focus is justified both by the problems of a mathematical nature that have been revealed by the research and by the discovery of a number of properties (mechanical, structural and associated with electrical conductivity) [1] that are distinctive of them. Configurations of minimal surfaces have been found in a wide variety of different systems: from the arrangement of calcite

G. Buratti (✉)
School of Design - Politecnico di Milano, Milan, Italy
e-mail: giorgio.buratti@polimi.it

M. Rossi and G. Buratti (eds.), *Computational Morphologies*,
https://doi.org/10.1007/978-3-319-60919-5_5

crystals that form the exoskeleton of certain organisms and the composition of human tissue to the basic structure of certain synthetic foams and the theories that explain the nature of astronomical phenomena. It is therefore understandable that leading scholars not only in mathematics, but also in physics, life sciences, material and structural science, medicine and more recently architecture and design, are studying them more and more intensively (Fig. 1).

A minimal surface is a surface whose mean curvature is always zero. This definition is closely related to Plateau Problem [2], also known as the first law: if a closed polygon or oblique plane (similar to a closed frame of any shape) is assigned, then there is always a system of surfaces, including all possible surfaces that touch the frame, which are able to minimise the area. In other words the problem is to identify the shape which covers the largest surface with the same perimeter.

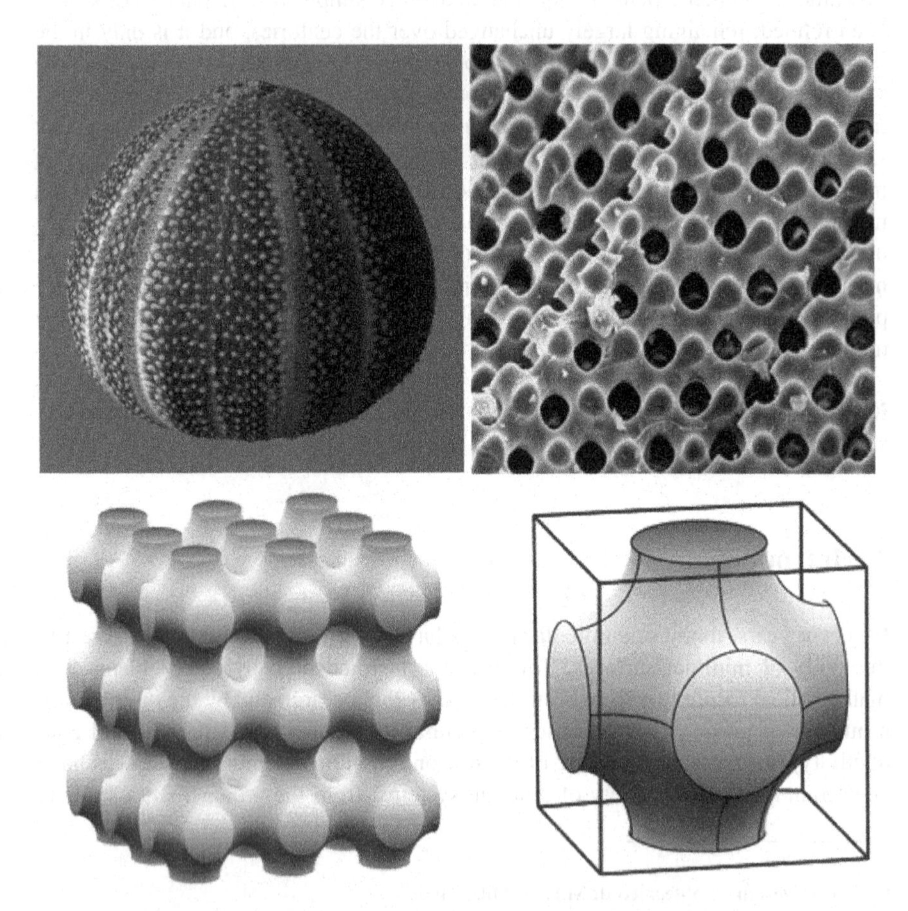

Fig. 1 Schematic of the primitive unit cell and periodic structure of Schwarz's P surface and cross section through a sea urchin skeletal plate showing resemblance to the P-surface

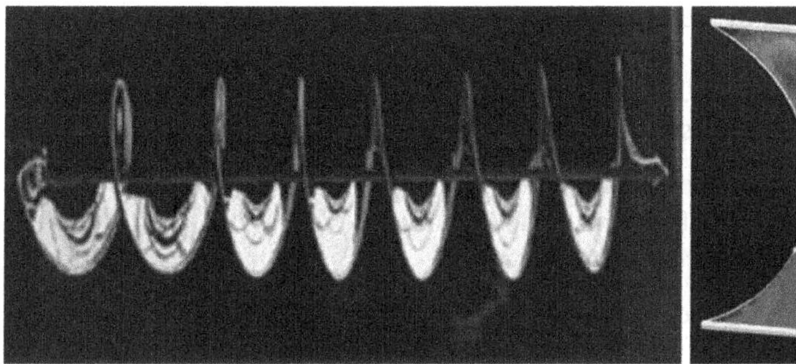
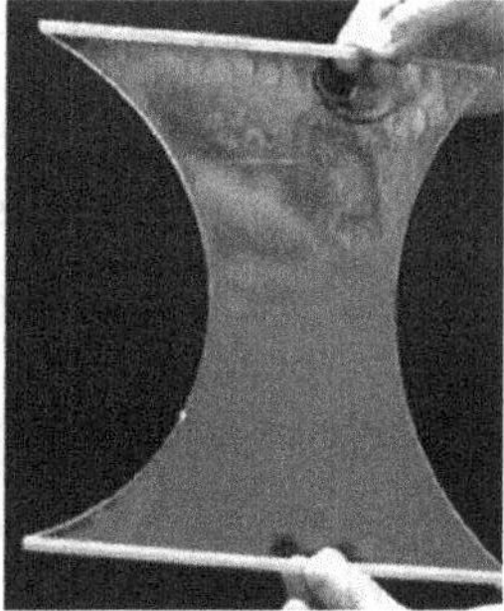

Fig. 2 The soapy lamina to forms a surface whose area is the minimum possible among those having the given contour. The phenomenon is well visible in *left figure*, in which is used a contour consists partly of flexible wires: the wires are markedly flexed inwardly from tension forces that attempt to reduce the surface of the lamina. In the *right image* a helicoid minimal surface formed by a soap film on a helical frame

Although it is true, as shown by Euler [3], that the property of minimising the area can be demonstrated only by a numerical formulation, it is only with Plateau and his observations on the behaviour of the laminae of soapy water that it has become possible to formulate a general principle that enables the creation of all minimal surfaces for which either the equations or the generating geometry are known. The experiments of Plateau consist of the construction of a kind of closed contour by using wire, with the sole condition that it circumscribes a limited portion of the surface and that it is compatible with the surface itself and its immersion in a soapy liquid. On the extraction of the frame from the fluid a set of soaped laminae is obtained that represent a portion of the surface under examination [4] (Fig. 2).

2 Triply Periodic Minimal Surfaces

In minimal surfaces family the Triply Periodic Minimal Surfaces (Tpms) are probably the ones that have the most interesting characteristics, including for project purposes. They are called periodic because they consist of a base unit that can be replicated, theoretically ad infinitum, in Cartesian space in three dimensions (triply), thus creating a new surface seamlessly and without intersections.

A uniform minimal surface is, usually, characterised by different curvatures; in other words, some surfaces are flatter than others. It follows that not all points of the surface support any concentrated loads equally well. If the same surface is, however, associated with a periodic distribution, i.e. the individual units are repeated next to each other, the physical iteration between the modules causes a compensatory effect that greatly increases their structural efficiency [5].

This is achieved, by the definition of minimal surface, through the use of as little material as possible. The advantages mentioned above are real when the surface obtained is a system under voltage or the material with which it is constructed is able to withstand tensile stresses and compression. In summary:

1. Tpms have natural geometric rigidity
2. Allow optimum use of materials
3. Configure stable and resistant structures

There is a large number of known embedded triply periodic minimal surfaces. Moreover, it seems that the examples come in 5-dimensional families, most of which are only partially understood. Our lack of knowledge of these surfaces makes it very hard to put them into categories. For the moment, we use the genus of the quotient by the largest lattice of orientation preserving translations as a guide [6]. In this paper we study three-periodic minimal surfaces that have three lattice vectors, i.e., they are invariant under translation along three independent directions. Numerous examples are known with cubic, tetragonal, rhombohedral, and orthorhombic symmetries [7].

The symmetries of a TPMS allow the surface to be constructed from a single asymmetric surface patch, which extends to the entire surface under the action of the symmetry group. The most important local symmetries of minimal surfaces are euclidean reflections (in mirror planes) and two-fold rotations [8] (Fig. 3).

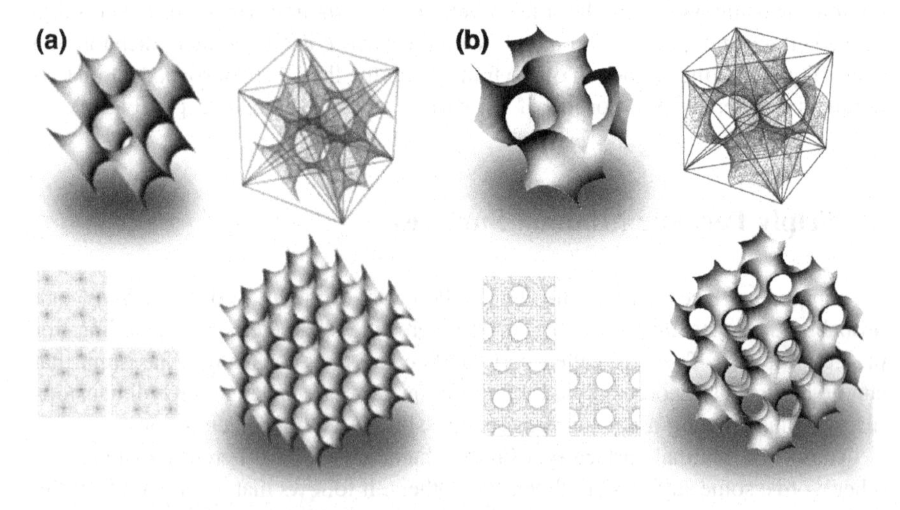

Fig. 3 From *Left*, one cubical unit cell, the fundamental quadrilateral and a four unit cubes in a triply periodic system. **a** Diamond minimal surface. **b** Gyroid minimal surface

3 The Case Study

First step is to find a way to manage the morphological complexity that characterizes the triply periodic minimal surface in a digital environment to study and eventually produce them. Using Grasshopper it was possible to write an algorithm that creates the minimal surface from its implicit description. The implicit method uses a linear function of three variables. Typically an implicit surface is defined by an equation of the form $f(x, y, z) = 0$. The algorithm translates the algebraic formula in a virtual geometric shape, which can be studied and manipulated (Fig. 4).

The process can be conceptually simplified thinking that the equation "choose", in a domain of points in Cartesian space, those belonging to the surface you select to represent.

The next instruction of algorithm is to connect these points creating the surface. Subsequently the surface is replicated to form the triply periodic minimal surface which will be adapted to the desired morphology.

This process allows a more thorough analysis of performance and it also permits to evaluated the artifact for the manufacture. A considerable level of complexity, could be analyzed via computer simulations, it's so possible investigates the mechanical properties of composite triply periodic minimal surfaces structures.

Mechanical characterization of the minimal surface was performed with uniaxial compressive simulation using COSMOS ITALIA (Cosmos Italia Srl. Cremona—Italy). Because of the complexity of the calculation it was necessary to simplify the geometry in order to obtain precise results. Therefore it is simulated the application of 1 kgf (10 N) on a iron parallelepiped of size $10 \times 10 \times 100$ mm and compared with two equivalents porous structures, composed respectively with the P surface and the Gyroid.

The digital simulation shows that even for very thin thickness (0.1 mm) the iron bar does not break, though it deforms considerably. This is because the stress,

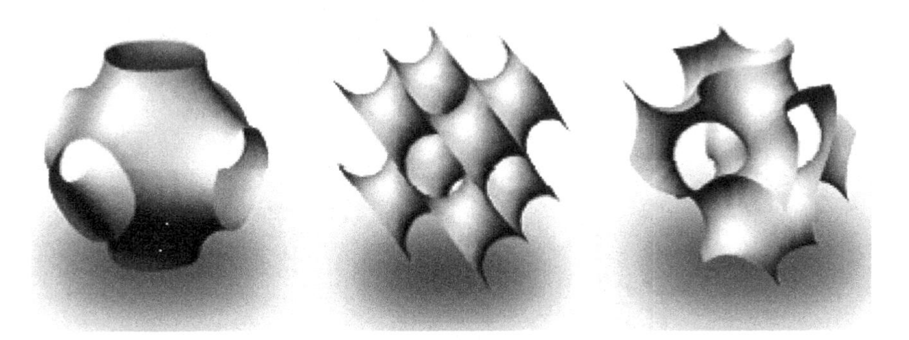

Fig. 4 From *left*: *P Surface* $\cos(x) + \cos(y) + \cos(z) = 0$; *D Surface* $\sin(x)\sin(y)\sin(z) + \sin(x)\cos(y)\cos(z) + \cos(x)\sin(y)\cos(z) + \cos(x)\cos(y)\sin(z) = 0$; *Gyroid* $\cos(x)\sin(y) + \cos(y)\sin(z) + \cos(z)\sin(x) = 0$

rather than focusing in a point, is distributed among all the single units that work as a unique element. We can say that the whole is greater than the sum of its parts (Fig. 5).

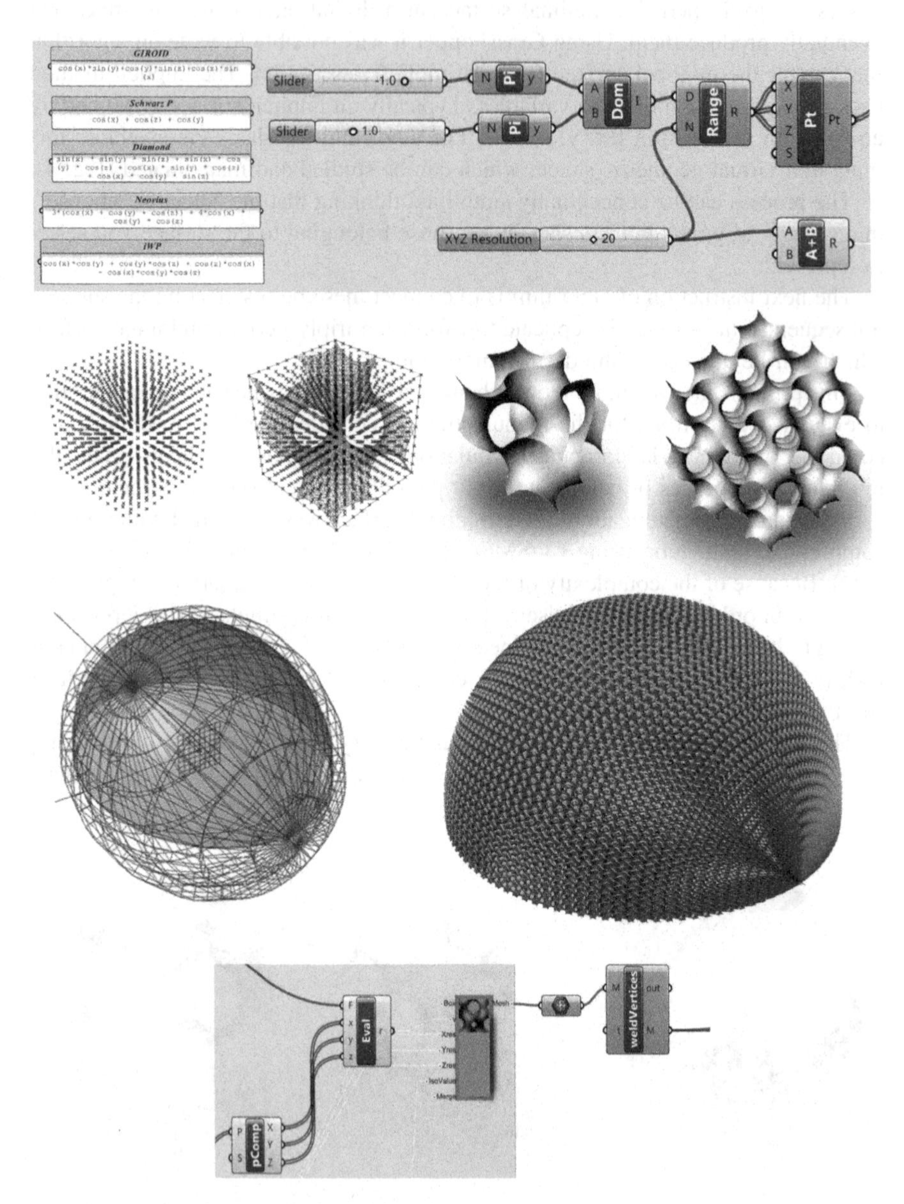

Fig. 5 a Equations are formulated in implicit formula, and collected in a library, than the Algorithm can describes different surface from the same points domain. The single minimal surface (a Gyroid) is placed in a periodic system. Subsequently the porous structure is used to create the desired morpholgy

If the thickness increases, the behavior changes significantly. For 2 mm of thickness deformation greatly decreases. P-surface bar has a displacement that is twice than solid bar, while Gyroid bar deforms even less. This means that if the solid iron bar deforms of 1 mm, that composed of P-surfaces deforms of 2 mm, while the Gyroid bar deforms of 1.6 mm. The most important aspect is that the two bars respectively weigh ten (P-surface) and eight (Gyroid) times less than the solid one, while maintaining good characteristics of stress resistance. This analysis demonstrates that minimal surface have significant potential: compared to conventional geometries the use of minimal surface maintains good mechanical properties with less weight and material (Fig. 6).

THICKNESS (mm)	DISPLACEMENT (mm)	STRESS (Kgf)	WEIGHT (g)
(a) 0,1	0,013	2,36	4
0,5	0,0012	0,273	19
1	0,0003	0,111	37
(b) 2	0,0001	0,044	74
Solid	0,00005	0,031	780

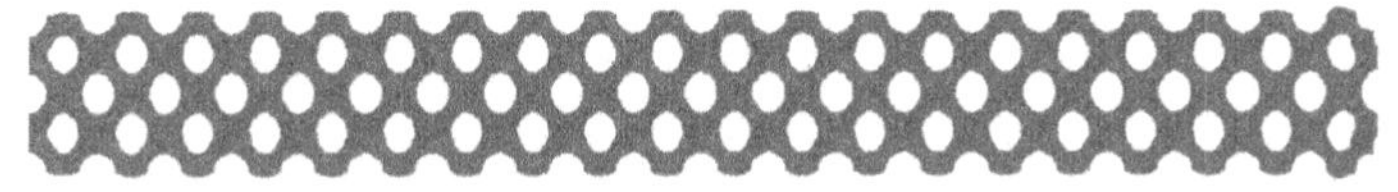

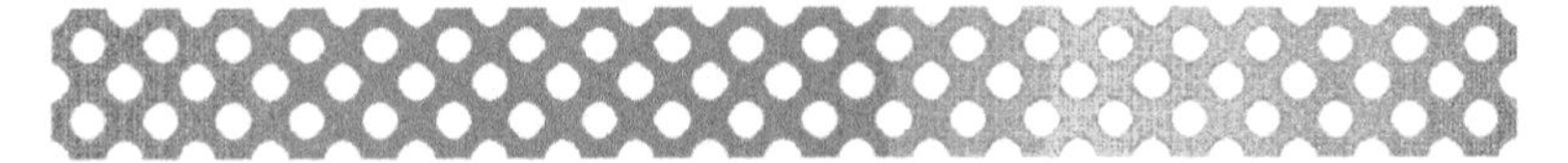

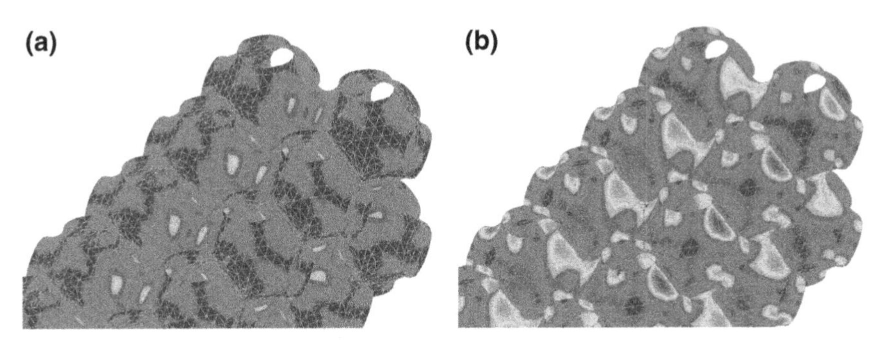

Fig. 6 Stress–strain behavior of a P-surface iron bar. **a** Stress 2.36 kgf, thickness 0.1. **b** Stress 2.36 kgf, thickness 2

References

1. Torquato, S., & Donev, A. (2004). Minimal surfaces and multifunctionality. *Journal of Applied Physics, 94*(5748), 5755.
2. Plateau, J. (1873). *Statique expérimetale et théorique des liquids soumis aux seules forces molèculaires*. Parigi: Gauthier-Villars.
3. Euler. (1952). *Methodus inveniendi lineas curvas maximi minimive proprietate guadentes sive solution problematic isoperimitreci latissimo sensu accepti*, in Opera omnia, a cura di C. Caratheòdory, Vol. 24, Fusli, Bernae.
4. Emmer, M. (2010). *Bolle di sapone, tra arte e matematica*. Torino: Bollati e Boringhieri editore.
5. Thompson, D. W. (1917). *On growth and form*. Cambridge University Press. Italian trasl: Universale Bollati Boringhieri, 1992.
6. Costa, C. (1982). Imersoes minimas completas em R3 de genero um e curvature total finite. Phd thesis, Rio de Janeiro, IMPA.
7. Brakke, K. (2013). *Triply periodic minimal surfaces*, Selinsgrove, PA: Susquehanna University. http://www.susqu.edu/facstaff/b/brakke/evolver/examples/periodic/periodic.html.
8. Horne, C. E. (2000). *Geometric symmetry in patterns and tilings. The textile Institute*. Cambridge: Woodhead Publishing.

Represented Models and Typological Algorithms: The Role of Parametric Models for the Design of Product

Andrea Casale and Michele Calvano

Abstract The design project in its evolution from idea to object requires investigations that make use of comparison among all models. The models, to be useful to the design process, must be easily editable in order to make quick assessments of the product that is being designed. The algorithm design, so far implicitly present in the mind of the designer, becomes a sort of conduct code of the product that can be explained by leaving some modification variables open. The conception of the behaviour becomes an integral part of the product and this is being resolved by the construction of a logical sequence with which compel the form to comply with the basic functional parameters, useful to classify the model within a "typological family".

Keywords Design · Generative · 3D models · Algorithmic · Computational

1 Introduction

In the famous 1996 book "Da cosa nasce cosa" ("From a thing was born a thing"), subtitled "Appunti per una metodologia progettuale" ("Notes for a design methodology"), Bruno Munari explores the mental and practical process that characterises the designer. He opens his analysis with a phrase taken from Archer: "[…] the problem of design was born out of a need". This represents the need to modify an object or a space, in order to respond, in the most suitable way, to the new demands that modern man experiences in his daily life. It is therefore the necessity of a more practical "container", which can be interpreted in two ways. Either as a settlement of the space, that depends on different activities which take place there, or as an entity to create artefacts suitable to solve specific social

A. Casale (✉) · M. Calvano (✉)
Faculty of Architecture Sapienza, Rome, Italy
e-mail: andrea.casale@uniroma1.it

M. Calvano
e-mail: michele.calvano@uniroma1.it

M. Rossi and G. Buratti (eds.), *Computational Morphologies*,
https://doi.org/10.1007/978-3-319-60919-5_6

demands of collective life. Munari, always in a clever humour, quotes the words of his friend Antonio Rebolini: "[…] when a problem cannot be solved, then it is not a problem. When a problem can be solved, then it is not a problem". These words describe the sagacity and multifaceted intelligence that has always characterized the planning activity of one of the most important figures of the Italian design, words that introduce the first and main section of his book: "Che cos'é un problema" ("What is a problem"). It is in this chapter that he distinguishes all steps-concepts that are necessary to switch from the problem to its solution: problem, problem definition, components of the problem, data collection, data analysis, creativity, materials and technology, experimentation, models, verification, and solution.

This list suggests a succession of stages, a path in which each of the identified moments is proposed in turn as a problem which, once is being solved, offers the opportunity to take the next step. In reality the design process is somewhat different. In fact, this process follows only part of this chain of conditions, which have a highly interactive role and constantly interact modifying each other reciprocally. The design idea can then propose choices that become part of the problem definition. The model can suggest solutions that can be creatively interpreted by the designer. The material interacts with the model. The verification acts directly with all the phases that precede it. Rather than a series of steps that lead to the solution of a problem, it is necessary to talk about a system democratically dialoguing, where each phase-problem has the right to speak, whose sole purpose is the only solution to the problem.

Munari, in his book, continues with a reflection on the relationship between designers and industry: "Molto spesso però l'industria tende a inventare falsi bisogni per poter produrre e vendere nuovi prodotti. In questo caso il designer non deve farsi coinvolgere in una operazione che è a solo profitto dell'industria e a danno del consumatore" ("Very often, however, the industry tends to invent false needs to be able to produce and sell new products. In this case the designer does not have to get involved in an operation which is only for industry profit and to the detriment of the consumer").

New information technologies, although more suited to handle the shape of the object, however, tend to remove this feature from the context, by altering the complexity of the artefact's design. The management of the features acquires a meaning more and more autonomous, expression of a creative and formal will, which lives a reality independent from the conditions that led to the problem. The informatics model, thanks to its realistic features, becomes the only moment of a phase (often only inventive) dangerously far from its final condition of real object, whose purpose is to respond to a specific function of social use.

A conceptual disconnection seems to occur between the criteria specific to the project and the form's management methods proposed by the new IT tools. Yet, in the process that leads to the problem's solution, it is perhaps possible to identify an algorithm process that is the basis of the computer method. Therefore, the succession of steps proposed by Munari could be understood as individual algorithms progressively arranged so that the solution of the first give useful information for the solution of the second and so on until you get to the final solution and thus to the project. This system seems to contradict what was stated previously: the project as a dialoguing democratically system where every stage actively interacts with all the others. It should therefore be imagined a system where each step is sufficiently binding and, at the same time, autonomous to allow direct action of the designer. Furthermore, the system should be able to verify how this action reverberates on the other levels, and then to the determination of the project.

Let's consider the example of a design problem, which investigates only the formal aspect of a chair's project. We know that the form is bound to certain dimensional requirements directly related to the physicality of the human body. While the seat should be slanted to prevent slippage and having a height from the ground that varies within a certain area; the backrest should accompany the back in an appropriate manner, so as to vary the height of the armrests within certain limits. These are just some of the ergonomic parameters that should guide the design of the chair. However, these parameters, although binding, can be modified in accordance with certain technical or formal decisions.

The parametric modelling systems seem to respond to these contemporary design requests. Nevertheless, even in this case we are witnessing two different conceptions that meet two specific requests. On the one hand, engineering systems where the design intervention is closely bound to a procedural hierarchy; and on the other hand, excessively free systems, fundamentally oriented to formal sculptural problem, where the constraints are arbitrary conditions imposed a priori. Therefore the need arises to identify a method that is able to re-establish that dialoguing process of the project design, able to identify problems and then manage the constraints actively and creatively, assuming them as essential, but always modifiable through the use of explicit algorithms.

2 From Logical Algorithms to Typological Algorithms

AAn algorithm is a formal procedure that solves a particular problem by a finite number of steps. A problem solved by an algorithm is said computable (Wikipedia). Extremely broad definition that gives an idea of how the algorithmic procedure is outside of specific conditions, but that represents any sequence of actions aimed at achieving a goal. The computer makes the procedure automatic through the use of specific syntax, that is, the programming languages. The designer is not a computer expert, but with the advent of visual programming (Visual Scripting) the compiling operation is extremely simpler.

Let's take, for example, a model created by extruding a generating curve along a directrix one (Fig. 1). This is a macro action that any 3D design software is able to perform in a few clicks, but which expresses a higher potential at the time when the action is divided into individual parameters and components (even of variable nature) that allow alteration and variation.

The case in the figure shows the extrusion of a circle along a straight segment. When constructing the cylindrical shape it has to be decided the initial arrangement of the circle, its radius and the extrusion direction. These data sorted in homogeneous lists may, at any time, be replaced with other one-dimensional forms and numeric data, elements suitably left open in the algorithm, and for this reason called "variables". Other variables may be the extrusion spatial angle and the length of the directrix line. Codes written in visual scripting make explicit the history of construction that emerges (Fig. 2).

By manipulating the data placed as editable values, the result you get is a changing, dynamic, interactive model. The model drawn on the computer becomes a kind of digital mock-up with which to test the way to communicate with the mental image to which the project aims. We can even go further deforming the generating curve and the directrix one within the definition. Only at this point the digital space becomes a virtual laboratory in which to experiment new forms (Fig. 3).

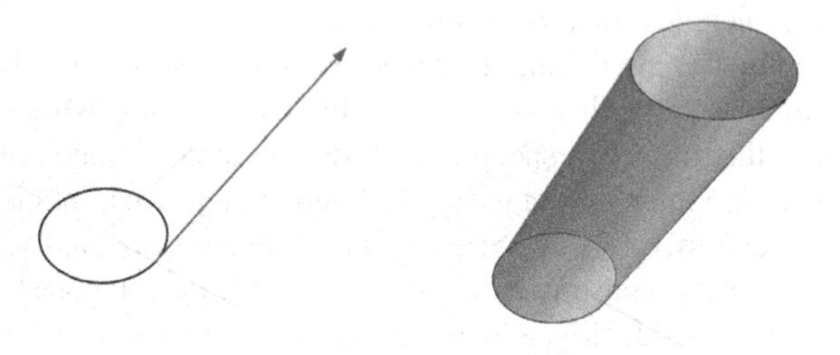

Fig. 1 Generating curve and directrix curve for the construction of a model of extrusion

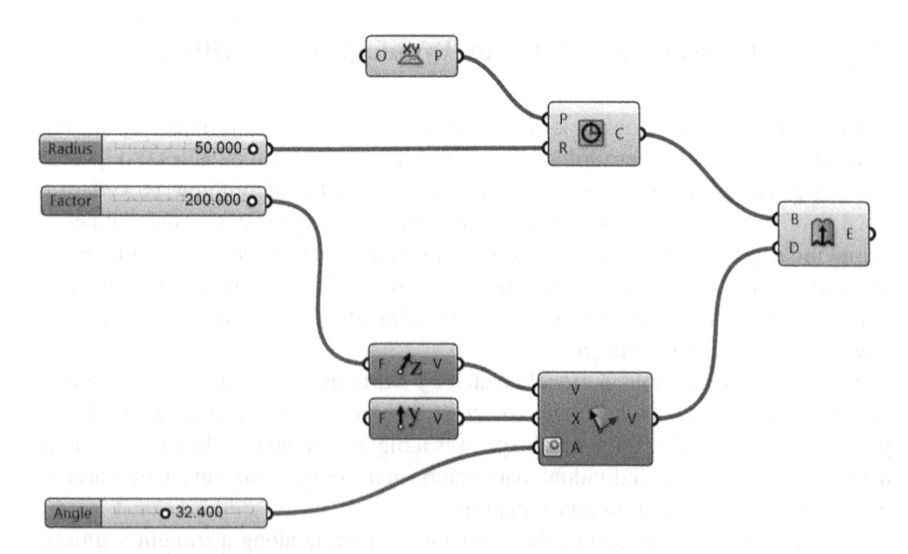

Fig. 2 Nodal definition which explicit in a computational algorithm a model of extrusion

The possibility of make algorithms explicit (through languages regulated by appropriate syntax) allows the designer to overcome the limit determined by the design software interfaces. In fact, these interfaces limit the digital space to be used simply as a huge drawing board. Accordingly, it is reaffirmed the usefulness of the logical algorithms' programming towards the design project. The idea, present in a conceptual manner in the designer's mind, must be split into elementary rules of construction linked by a sequence of actions aimed at the development of a code which makes it computable. Different outputs correspond to each value of the input variables. These output then result to be mutations of the same project tied together by a common code. This connection ensures a kind of relationship between the models that are generated by scrolling through the input data. Furthermore, by varying some inputs it is possible to obtain different forms and models linked to a common definition. These models belong to the same family, which is the function of a common generative structure that guarantees a degree of relationship. In this case, the mutation of the variables involves different models (though with a common genetic) that allow the designer to choose the form that best meets his needs, or the ability to create a family of forms that satisfy a homogeneous category of functions (Fig. 4).

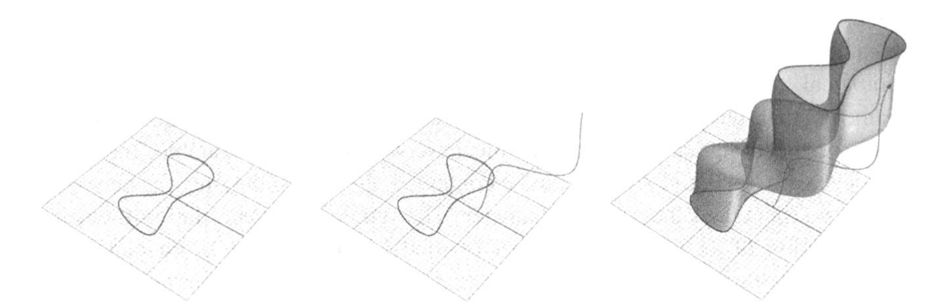

Fig. 3 Mutation of the directrix and generating curves used as variables in the definition in Fig. 2

Fig. 4 Family of forms that satisfy a homogeneous category of functions regulated by the same typological definition

In this variable models' system you can still find invariant elements evident in the form, but dependent on the common algorithmic definition that generated them. For this reason the definition that creates these models is called "typological algorithm".

The definition acquires greater recognition if we think about the meaning of the word "type" in a non-specialized language: we see that it is equivalent to a general form or a set of properties that are common to a number of individuals or objects. Type is synonymous with class, family, gender; therefore a category that results from the classification on a set of objects. The typological algorithm, like the type, dos not have an objectual nature, but on the contrary it is characterised by a conceptual essence: it brings together a family of objects that possess all the same essential conditions and live in common ranges. Belonging to the same typological algorithm involves the recognition of functional and formal common variable parts.

3 The Case Study

Here are a few experiments conducted on a typological algorithm with which it is possible tHere are a few experiments conducted on a typological algorithm with which it is possible to create the bottle's models as shown in the Fig. 5. The typological algorithm is being composed keeping in mind some functional parameters that must be guaranteed in the different models that will be generated by means of the designed definition. In the described case, for example, the fixed parameters are the liquid's volume to be contained, the size of the ring, the height of the entire object, and the support diameter. The typological algorithm will ensure the generation of an undefined number of models, always able to guarantee the presupposed characteristics. We have to start from the concept of primitive form that subtends the generated one, which is the result of the relationship between open longitudinal curves and closed transverse ones. The examined type, namely the bottle, can be drawn either by means of the revolution of the generating curve along the directrix one, or through the rolling with a surface of a closed curves' sequence by using the loft tool. As can be seen in the figure, the complexity and generativity of the form is a function of the variability of the curves to be rolled with the surface.

Now let's try to describe the typological definition below (Fig. 6). The first part of programming is devoted to the identification of the model's height [1] and the number of circumferences to be rolled [2]. These circumferences have variable radius but are coordinated by a Bezier function [3] for a controlled variance. In order to make the morphology of the model more interesting, radial deformations were introduced on the circumferences by manipulating the NURBS' mathematics that represent them. The reconstruction of the curves allows the addition of a greater number of control points [4] that, once moved radially [5], produce transverse articulated outlines. The outlines are further subjected to an incremental rotation [6] from the bottom upwards in order to simulate a twist once that the curves are being rolled by the surface [7]. This typological definition generates models variously

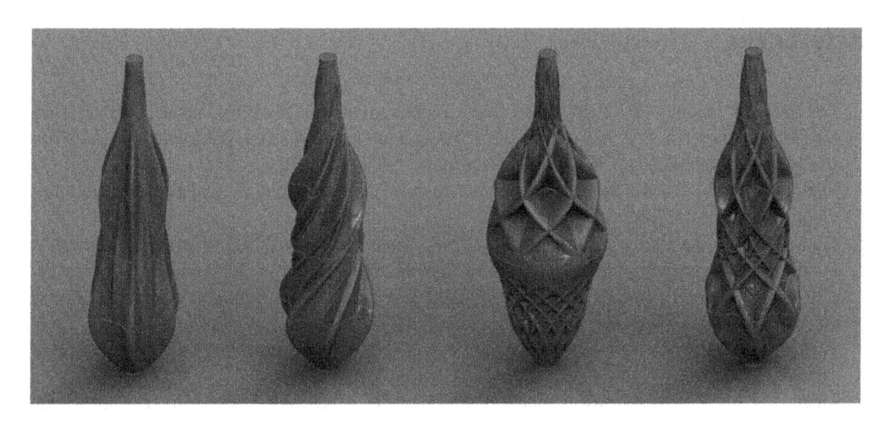

Fig. 5 Experimentation of a typological definition capable of creating families of bottles

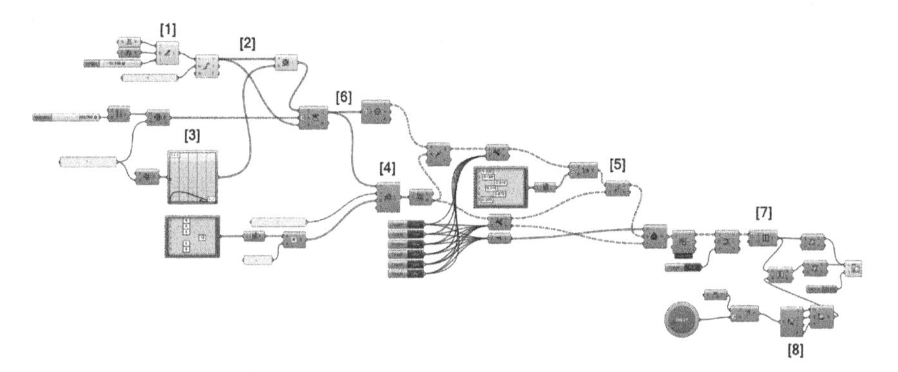

Fig. 6 Typological algorithm capable of generating a family of models with the same variable parameters

tapered: longitudinal streaks with variable overhang; or encircling streaks with variable overhang and a further indefinite number of models connected to those described above, but variable in relation to the open parameters of the delineated definition. The nodal programming is further articulable by adding a mirror plane passing through the revolution axis of the model with the possibility to rotate around it [8].

At this point, we can conclude that the definition opens a new window between the infinity of the models. This further amendment adds new variables to the form, giving the possibility to catalogue the just-changed definition as a new typological algorithm, distinct from the previous one.

References

1. Eck, M., & Hoppe, H. (1996). Automatic reconstruction of B-spline surfaces of arbitrary topological type. In *Proceedings of the 23rd annual conference on computer graphics and interactive techniques* (pp. 325–333). New York, NY: ACM Press.
2. Dinh, H. Q. (2000). *A sampling of surface reconstruction technique.* College of Computing Georgia Institute of Technology.
3. Gerbino, S., Martorelli, M., Nigrelli, V., & Speranza, D. (2002). Reverse engineering techniques in improving the development phases of a new footwear. In *Proceedings of the 3rd CIRP international conference on intelligent computation in manufacturing engineering (ICME)* (pp. 395–400), Ischia, Italy.
4. Brevi, F. (2004). *Il Design delle Superfici. I modelli digitali per il disegno industriale*, ed. Poli. Design, Milano.
5. Graziano, M. V. (2004). *MI Il modello integrato*. In MIGLIARI, Riccardo. Disegno come Modello, Kappa (pp. 59–62).
6. ISSA R. (2009). *Essential mathematics for compute national design.* Robert McNneel & Associates.

Project Rule-Checking for Enhancing Workers Safety in Preserving Heritage Building

Antonio Fioravanti, Francesco Livio Rossini and Armando Trento

Abstract Historical building refurbishments as a whole and rich environment can be affected by many variations, due to non-calculated risks and hidden problems during construction, so that the original projects are distorted. These risks involve many problems: time lengthening, cost increase and, when conditions are unfavourable, the possible exposure of workers to the probability of having accidents. This study, in the early design phase, focused on tools and methods able to reveal interferences, allowing them to be solved in the design phase. Risk evaluation is a fundamental procedure that allows actors, whoever they are (stakeholders, planners, decision-makers and implementers), to identify measures of prevention/mitigation and to schedule implementation, improvement and control. We propose to demonstrate how the risk, declined in the main aspects that affect the construction industry, can be modelled properly and reduced thanks to the innovation of existing design methods and tools. Particularly, we maintain that tool transformation, often related to a profound change of methods and vice versa, may lead to the study of the relationships between design, building and activities, collaboratively including actors in the design process. To cope with complex problems, BIM models should be able to implement and manipulate multiple sets of entities, qualified by clearly established relationships, belonging to comprehensive structured and oriented (sub-) systems. Given the state of the art and the potential not yet fully expressed by 4/5D BIM systems in the field of Project Management (PM), this paper reports on on-going research oriented to formalise Design knowledge based on field experience, made possible by a particularly positive convergence where the authors have been contractually designated by an international private company well known in

A. Fioravanti (✉) · F.L. Rossini · A. Trento
Faculty of Civil and Industrial Engineering, Sapienza University of Rome,
Via Eudossiana 18, 00184 Rome, Italy
e-mail: antonio.fioravanti@uniroma1.it

F.L. Rossini
e-mail: francesco.rossini@uniroma1.it

A. Trento
e-mail: armando.trento@uniroma1.it

M. Rossi and G. Buratti (eds.), *Computational Morphologies*,
https://doi.org/10.1007/978-3-319-60919-5_7

the fashion sector for leading coordination among different specialists in the challenging functional restoration works of a significant historical building, known as "Palazzo della Civiltà Italiana", also known as "Square Coliseum". To arrive at a real integration among the architectural design solutions of the project, the protection of the listed building and the implementation of safety measures, a methodology has been developed that allows a continuous exchange/upgrade of information among these entities. The operative mid-term results can be subdivided into two levels: (1) one, more pragmatic, has been to boost workers' education about non-standard operative tasks by means of accurate construction narrative visualization; (2) another one, more theoretically, oriented towards model "judgment-based" rules aimed at supporting automated reasoning in Safety Costs' evaluation and assessment.

Keywords BIM models · Risk evaluation · Design process

1 Introduction

Design is an activity related to the future, aimed to define in advance forms, meanings and performances. But often, the future orientation of this activity has to be compared with existing buildings, in which we can find some constraints–in the first instance, the respect for historical features–which guide the choices of designers and the phasing of construction.

Specifically, in this huge field of research, only a few researchers systematically investigate how to integrate the design solution aimed at heritage preservation with the Health and Safety (HS) measures, enhancing the effectiveness of these measures and, with appropriate workflows and tools, a predictive model of how to enhance construction performance mitigating risks.

In A/E/C fields, risks are various, involving economic, legal and health aspects. This paper focuses on the work-related risks like accidents and injuries, considering that the building industry is one of the most hazardous activities in the European Union, where the fatal accident rate is approximately 13 workers per 100,000 against 5 per 100,000 on average for other sectors [1]. Italy has the highest rate of workplace accidents and deaths in the EU, where the estimated high number of accidents, which does not include the unfortunately large percentage of undeclared workers. These data reveal how difficult but important it is to prevent hazards that can affect the building process. These difficulties are more evident in the case of heritage building processes, where it is more complicated to connect actors (workers), activities and places, considering that the preservation of the historical object often does not allow the installation of safety measure like, for instance, permanent parapets, the connection between historical walls and safety ropes, temporary installation of the site construction offices and hygienic services.

Furthermore, as well as these material difficulties we have to consider immaterial issues; according to Ciribini [2], the poor performance and the lack of managerial

Fig. 1 Construction site of the case study—"Square Coliseum"—under construction. Archivio Storico ente EUR S.p.A.

skills affecting the construction sites are due to a lack of co-operation among the various actors such as clients, organizations, main contractors, subcontractors and supplies. Moreover, the current economic situation is worsening the trend, and companies are forced to work with low profit margins, trying to make savings in all aspects of the construction and neglecting safety. On the other hand, designers, engineers and contractors have an important influence on the health and safety of building site employees. In fact, in current practice, there is a dangerous tendency to underestimate the importance of planning safety during construction, considering this activity as a bureaucratic exercise or, at worst, a matter to be entrusted to the experience and habits of individual workers involved in the process as it can be seen in traditional construction (Fig. 1) where the main part of the working phases occurs following traditional schemes and habits.

Conversely, with regard to cases of high complexity and costs, the need to integrate the design solution with safety planning measures is realised in only a few cases. Only sporadically are advanced ICT automated techniques employed to support the heterogeneous competences and duties demanded by the professional profile of the Health and Safety Coordinator (HSC).

2 Background

2.1 Policy Framework and Current Practise

Since the adoption of the EU Directive (89/391/EEC) and the related Construction Site Directive (CSD–91/57/EEC), European building designers have been legally required to consider health and safety in their projects. However, previous studies have shown that designers in general –not just in the construction industry-fall short of satisfying this obligation [3] mainly because most contractors often fail to implement their health and safety plans [4].

In addition, the importance of the economic evaluation of safety is often underestimated: international scientific literature demonstrates that it frequently happens that HSC, because of contextual construction planning uncertainty, ends up by identifying general solutions that fit loosely, without considering any alternative at a comparable cost even when it affords an improvement in safety measures. This state of mind implicates a sure combination between designers and executors, who are induced into erroneous Safety evaluation and Estimation determining an inefficient use of resources.

One important issue is also the lack of interest shown by customer, companies and designer in starting to plan safety and construction phases at the early project stages, before the solutions are applied. This habit actually leads them to serially reuse the same strategy, not considering that each site is a unique prototype in itself. This is even more dangerous considering that text, charts and evocative pictures, produced by similar old construction sites, are not sufficient for the purpose of predicting when and where accidents could occur. In recent decades, this erroneous traditional way has slowly been supported by information technology enabled approaches for construction safety using virtual design and simulation of construction operations.

2.2 *BIM Methodology in Health and Safety: State of the Art*

In fact, in most advanced economies across the globe, designers have started to use Building Information Modelling (BIM) sharing databases for synchronized use between companies, designers and inspectors. Over the years, BIM-based software packages have established firm positions, and are used by AEC-FM professionals; the most important BIM software features 3D/4D safety planning, management and communication including 3D modelling and viewing capabilities, 4D tools and features, tools for analysing risks or safety of the design and plans and data exchange capabilities. Commercial IFC-compliant BIM models allow software interoperability, phase-dependent project analysis and clash detection among components.

Recently, in addition to the above capabilities, new platforms have been introduced to define and manage 4D models. These new environments are focused on assisting Construction Managers in describing the sequence of operations to be performed by linking a narrative timeline to building entities.

But currently, in today's A/E/C safety planning and management practice, what can be observed in terms of most common application methods is the parallel use of various, seldom compliant and interoperable software. For instance, site layout drawings or specific illustration of tasks to be performed are two-dimensional representations, only implicitly linked for the sake of analogy to the description of risks, hazards and safety prescriptions of what is expected to happen in reality. Moreover, with regard to descriptive documents, the use of spreadsheets or pre-designed forms for facilitating data entry procedures, are not adequately supplied by, or linked to, a detailed library and or structured databases.

In order to inform different domain actors, eliminate hazard and reduce risks, several companies and organisations focused on BIM are nowadays researching the implementation of, and the connection to, tools oriented towards workers' safety training and educations, design for safety, safety planning, (job hazard analysis and pre-task planning) accident investigation, and facility and maintenance phase safety.

2.3 Efforts to Adopt BIM in Health and Safety Management

Zhang et al. [5] present a system that automatically analyses a building model to detect safety hazards and suggest preventive measures to users using pre-defined algorithms. Within this framework, a rule-based engine is implemented on top of a commercially available BIM platform to show the feasibility of the approach. As a result, the automated safety checking platform developed informs construction engineers and managers by reporting why, where, when, and what safety measures are needed to prevent fall related accidents before construction starts.

Furthermore, Bansal [6] studies the benefits of translating 2D drawings into a 3D model, which allows its components to be mentally linked up to the respective activities defined in the schedule, in order to understand the execution sequence during safety planning. In this way, sequence interpretation and hazard identification vary with the level of experience, knowledge and individual perspective of the safety planner. This system works by means of four dimensional (4D) modelling or building information modelling (BIM), in order to create the simulation of the construction process by linking execution schedule with the 3D model.

Finally, during safety review, the process is planned in terms of sequence results from a hazardous situation, eventually corrected within the GIS itself before actual implementation.

Moreover, this workflow determines a synchronised database, useful for manual procedure of using BIM technology for safety planning: VVT visualized BIM-based 4D safety railings for fall/edge protection in Tekla Structures. However, most of the existing efforts in safety planning either largely rely on human input or offer knowledge-based/semi-automated implementation.

Exploring how hazards, risks and solutions could be built into BIM, activating automated reasoning in a way that designers, constructors and users find useful, is an urgent task to pursue in international research.

2.4 *Safety Requirements and BIM: Potential and Limits*

Since rules, guidelines and best practices already exist, they can be used in conjunction with existing 3D design and schedule information to formulate an automated safety-rule checking system (Figs. 2, 3). Safety conditions come to light and are thus resolved within the construction process, as the project progresses.

Compared to existing BIM applications such as clash detection and BIM-based quantity take-off, a basic requirement for a rule-based checking system is that each building object carries information: for example, object name, type, attributes, relationships and metadata including object identification (ID) number, date, and author creating model elements. It is important to point out that the schedule data need to be linked to the building object data since the assigned protective system needs to be updated accordingly.

There are rule-based platforms available that can apply rules to IFC building model data. The Solibri Model Checker [7] in particular is a software package that can be used to combine the viewing and the examination of the content of various BIM-models. This software provides special tools for rule-based automated checking and analysis as well as quantity and other data take-off. Users can also edit

Fig. 2 The BIM model of interior design of "Palazzo della Civiltà italiana": construction of an office-module. Photo by authors

Fig. 3 The actual condition in the construction site compared to the BIM model in Fig. 2. Photo by authors, May 2015

and create new rule sets. However, effective automation cannot be enhanced until the design/management tool can rely only on entities formalised and structured "per se" and fulfilled by isolated information. BIM models should be able to implement and manipulate multiple sets of entities, qualified by established relationships, belonging to organically oriented (sub-) systems. More sophisticated information modelling structures are needed in order to allow querying and computing design/Construction/Safety entities at a higher level of abstraction. Even if the 3D/4D BIM approach is certainly more advanced and valuable than methods and tools traditionally used by Health & Safety coordinators, it is affected by some critical issues:

- Classes and detail accuracy of existing parametric entities are oriented mainly toward Building design and less toward Construction. Construction and Safety are still on the fringes of the world of BIM;
- Software management information that is not structured enough to allow model assembly relations among entities (e.g. hierarchical or topological) or—very critical themes in AEC practice between building entities like components, spaces, equipment, machines and actors;
- Consequently, 4D BIMs support a low effectiveness level of rule modelling, with limited possibilities of rule based automatic checking, and still far from "judgement-based" reasoning. It is believed that the area of real time project information on safety risk and therefore measures to be taken by the workers, is still a very urgent and open research problem.

3 Framework and Methodology

The overall research aim is to develop a method for enhancing the quality of modelled/managed information and defining an implementation path toward a desirable dynamic and holistic knowledge-based support system. This paper reports on a mid-term research work aimed to define and implement a systematic methodology for assisting HSC activities, aimed at calculating the safety- related performance of a project and providing a consistent basis for comparisons between different safety design solutions. The main objective can be subdivided into two levels:

- The first is targeted at educating workers regarding non-standard operative tasks;
- Another one, more theoretically complex, oriented to model "judgement-based" rules,

aimed at supporting automated reasoning in Safety Costs' evaluation and estimation. On one side, a 4D BIM-based site plan is being modelled and used to produce illustrative representations of the site organisation and of the safety arrangements. The 3D views together with 4D narrative can be used for the orientation of site workers, task guidance and instructions, for informing about risks. On the other side, to manage safety solutions, both predicting during the building design phase and evaluating–in real time-during the construction phase, we formalise a few representative rules to be checked by the system and some related operative actions to be suggested as a consequence. In order to have automatic judgement, as a starting task a clear definition of rules to be checked is needed. Specifically, in the safety design phase we are trying to address the prevention of hazards as a case study. According to the Italian D. Lgs. 81/2008, inheriting EU Directive n. 92/57/EEC, safety risk assessments are a function of physical entities on one side and, on the other side, can be linked to more abstract coordination process activities. Primarily, we analysed how to represent a safety condition in terms of required objects, attributes and relations. We then formally designed a routine to run the risk evaluation, intended as complex rules to be applied to the objects, attributes and relations involved. The formal representation of safety categories and rules to be applied have been implemented using a compact knowledge representation structure and managed by ontology-based technology. In order to assist human decision makers in safety planning and scheduling activities, the outcome of the hazard prediction.

4 Enhancing the Representation of Safety Entities and Rules

In order to have automatic judgement, as a starting task, a clear definition of rules to be checked is needed. Specifically, in the safety design phase we are endeavouring to address the prevention of hazard as a case study. According to the Italian D.Lgs.

EU Directive n. 92/57/EEC, safety risk assessments are a function of physical entities on one side and, on the other side, they can be linked to more abstract coordination process activities.

Primarily, we analysed how to represent a safety condition in terms of required objects, attributes and relations. We then formally designed a routine to run the risk evaluation, intended as complex rules to be applied to the objects, attributes and relations involved. The formal representation of safety categories and rules to be applied has been implemented using a compact knowledge representation structure and managed by ontology-based technology.

In order to assist human decision makers in safety planning and scheduling activities, the outcome of the hazard prediction assessment is an automatically generated alert which would inform the necessary parties. In order to avoid a typical limitation in the evaluation of health and safety risks, by means of the previously presented knowledge modelling approach, authors implement some classes of a more general Health and Safety design ontology. "Judgement based" evaluation can be performed by using quantitative indicators formalised by Gangolells et al. [8] which are based on data available in the project documents/models. Indicators measure physical property values of construction/safety related entities not available in current BIM/IFC models. The implementation steps are the following:

- Represent Safety Risks related to Construction Activity (e.g. expressed in OWL language by means of ontology editors, e.g. Protégé);
- Represent an extended library within BIM of construction/safety related entities (by means of API), with special emphasis on Risk Indicator physical properties;
- Link Construction Activity (time/space instantiated) with actual BIM/IFC safety related entity (e.g. in Autodesk Navisworks). After the previous steps have been successfully implemented, the system will assist safety designers and HS decision-makers by supporting them automatically, calculating the Risk indicator of the instantiated project Construction Activities.

The proposed ontology will allow the classification of all terms (aspects, impacts, risks, and procedures) related to the Health and Safety "Judgement- based" evaluation as well as the relationship that exists among sets of design objects. Then again, each class will be enriched with different properties which will be used by the decision-making tool to identify the main significant Health and Safety conditions in each design/construction process and, moreover, to evaluate their impact in a specific construction project in order to provide procedures.

5 Case Study: The Bridge into the Cloister of "Palazzo Della Civiltà Italiana"

This paper reports on on-going research aimed to formalise Safety design knowledge based on field experience. The research is made possible by a particularly positive convergence. In fact, the authors have been contractually designated by

Fig. 4 The bridge during construction, July 2015. Photo by the authors

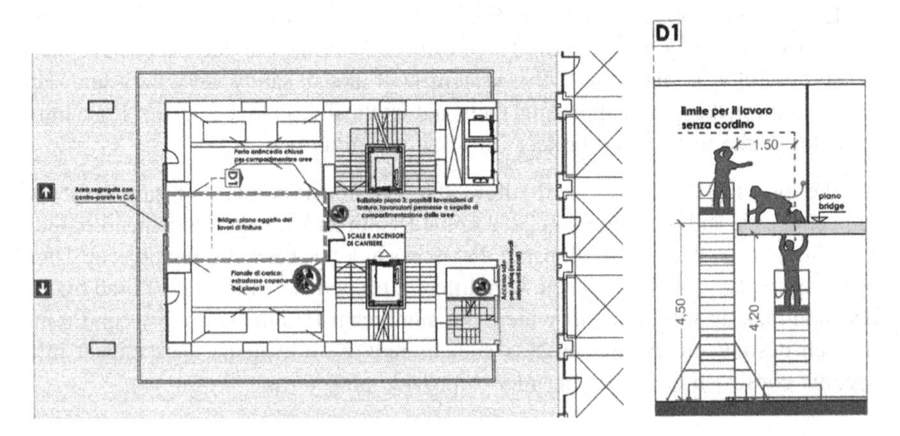

Fig. 5 Detailed design of the Heath and Safety construction procedures. Photo by the authors, July 2015

Fendi S.r.l., an international private company well known in fashion sector, for leading Health and Safety Coordination in the challenging functional restoration work of a significant historical building the "Palazzo della Civiltà italiana" (Fig. 6) better known as the "Square Coliseum", where the company intends to move its

Fig. 6 The "Palazzo della Civiltà italiana", Rome. Creative Common (CC-BY-SA-3.0)

headquarters. This public property is an iconic building of particular historical interest and is listed by the Minister of Cultural Heritage. This huge building, constructed between 1938 and 1943 was designed by the architects Guerrini, La Padula and Romano, and represents one of the most typical examples in the modern history of architecture of the Italian classical trend towards rationalism. The main objective of this sophisticated intervention is to ensure, as a function of the triad Time–Cost–Performance, a product of high architectural quality, performed in compliance with the history of the building, planning the work phases in such a way as to bring them closer to serial standards of production. The most difficult intervention in this context, was to build the bridge, located on the third floor, that connects the staircases and elevator zone directly with the Chief Executive Officer's (CEO) office (Fig. 4). This steel structure, suspended in this evocative space, is represented by a system of beams that create the minimum visive impact, creating a relaxed zone suspended in space.

6 Conclusion

Safety planning can be a part of 4D production planning. This creates safety planning practice that is undertaken earlier than is traditionally the case in construction projects, and furthermore can ensure a more detailed planning level. A BIM- based site plan can be used to produce illustrative representations of the site and safety arrangements, and the views can be used for the orientation of site workers (Fig. 5), task guidance and instructions, informing about the risks. Safety control and evaluation is based on both proactive and reactive performance indictors relying on a percentage of safe work packages and actual accident data. The proposed methodology and implementation path is aimed to support designers and construction planners in visualizing and evaluating safety risks, related to construction activities by means of risk indicators (Fig. 6).

Vice versa, safety designers need to spend more resources in order to utilize the construction work breakdown structure and thus enhance the geometric model with construction safety related entities by including more details which will result in an ever-changing work site schedule.

7 Future Works

In summary, the expected general research results are:

1. To enhance safety related details of the BIM which is typically modelled by the project designers. Specifically: accuracy of construction site entities and building construction sub/phases linked to Safety work breakdown structure;
2. A reusable safety knowledge model introducing a system of hierarchical relationships in time and space to represent parametrical Safety conditions.
3. To establish relationships between activities and tasks and a hierarchical relationship between operators and devices and/or safety equipment;
4. To predict site hazards and define corrective measures by means of a stored library populated by alternative resolutions. To support HS design by proposing realistic solutions to resolve identified issues.

References

1. Aulin, R., & Capone, P. (2010). The role of Health and Safety coordinator in Sweden and Italy Construction industry. In *18th CIB world building congress, W099—Safety and Health in construction* (pp. 93–96), The Lowry, Salford quays, UK.
2. Ciribini, A. L. C. (2013). *L'information modeling e il mondo delle costruzioni,* Sant'Arcangelo di Romagna (RI), Maggioli editore, 1a edizione, ISBN 8838761574.
3. Gambatese, J., Behm, M., & Hinze, J. (2005). Viability of designgin for construction worker safety. *Journal of Construction Engineering and Management, 131*(9), 1029–1036.
4. Swuste, P., Frijters, A., & Guldenmund, F. (2012). Is it possible to influence safety in the building sector? A literature review from 1980 until the present. *Safety Science, 50*(5).
5. Zhang, S., Teizer, J., Lee, J., Eastman, C. M., & Venugopal, M. (2013). Building information modeling (BIM) and safety: Automatic safety checking of construction models and schedules. *Automation in Construction, 29,* 183–195.
6. Bansal, V. K. (2011). Application of geographic information system in construction safety planning. *International Journal of Project Management, 29*(1), 66–77.

7. Solibri model checker, http://www.solibri.com/products/solibri-model-checker/. Retrieved from November 13, 2015.
8. Gangolells, M., Casals, M., Forcada, N., Roca, X., Fuertes, A., Macarulla, M., et al. Identifying potential health and safety risks at the pre-construction stage. In *Proceedings of the 18th CIB world building congress W099* (pp. 59–73), Salford, UK.

Part II
Design and Responsivity

Architectural Templates: A Hands-On Approach to Responsive Morphologies

Attilio Nebuloni and Giorgio Vignati

Abstract The ability to acknowledge external inputs in order to make them part of its organizational structure is one of the aspects of "non-linear" architecture that today is perhaps more peculiar to a new design trend, whose main features lead to the definition of complex and changing space systems. This implies the adoption of design tools and techniques capable of dynamicity. Moreover, thinking in terms of architectural competence and behaviour, in spite of its own image, means to adopt in design the logic of coding and computational. This paper discusses the background of responsiveness, its relation to architecture dealing with parameter's value for instability and dynamism, and explain computational strategies and design methods of a hands-on application based on a simple origami structure, which is also finalizing to design an architectural template for both testing and designing kinematic architectural components, as the main output of the research itself.

Keywords Computational design · Responsiveness · Adaptive architecture · Physical computing · Origami architecture

1 Background

The relationship between design and responsiveness is the focus of a line of architectural research with a focus on understanding the composition of methods and forms deriving from nature.

Created in the technological context of the architecture of interaction, the current interest in responsive design has common roots in the "*environmental*" interpretation of design, which targets context as a complex system to be addressed.

A. Nebuloni (✉) · G. Vignati
Department of Design, School of Architecture Urban Planning Construction Engineering, Politecnico di Milano, Milan, Italy
e-mail: attilio.nebuloni@polimi.it

G. Vignati
e-mail: giorgio.vignati@polimi.it

M. Rossi and G. Buratti (eds.), *Computational Morphologies*,
https://doi.org/10.1007/978-3-319-60919-5_8

This does not necessarily mean a focus on "*green*" or environmentally responsible forms of architecture, but rather on the simpler concept of designing systems, forms and components capable of interacting and combining symbiotically with a context.

While sharing a number of common elements, the definition of the concept of responsive architecture, supported by the current interest in the theme in much international architectural study, is not in itself univocal. On the contrary, in the current hybrid dimension of design we may find a plurality of different interpretations which differ in terms of the specific context to which the concept is applied. These include Michael Fox and Kas Oosterhuis's definitions, the synthesis of which significantly frames particular features and contexts of the relationship between interaction and architecture. Fox says that "*interactive architecture can be defined as the total integration of the disciplines of interaction design and architecture. [...] if architecture is to continue to respond to the possibilities of technological innovation that surround it as profession, then we may no longer ask "What that build?" or "How was it made?*", but rather "*What does that building do?*". And Oosterhuis, in framing the context of his research, states that "I*nteractive architecture is about the potential for digital systems to make decisions about our living environment and then influence that environment*" [4, 10].

In brief, the term "*responsive architecture*" may be defined as the capacity of a structure or system to receive external input—from the environment or society—as the principal material of its formal organisation. Relational capacity is therefore an internal property and a key goal of such organisations. Compared to systems founded solely on the physicality of form, these organisations are therefore more interested in the diagrammatic aspect and the relational organisation of their structures, which may be observed in terms of their capacity for "*dialogue*" and interaction with the open, constantly changing dimension of the environment in which they fit. This aspect is not so common in the domain of architecture, if observed, for example, through the evergreen filter of Vitruvius' firmitas, which does not contemplate the possibility that architecture may change over time.

The relationship between interactivity and architecture summed up in the concept of responsive architecture therefore applies to a more general aspect of architecture targeting the modes and forms of its organisation and the tools used by architects and users to mould and modify their environment, and therefore the methods and tools used to ensure dialogue and discussion between people and the places they dwell in.

The result is a new idea of architecture which is no longer based on the principles of solidity and unchangeability, but open and ready to accept change. A form of architecture based on a process, in which the result of its application in space does not depend on the dynamic behaviour of its users and, at the same time, the changing needs and specific features of the context in which it arises. A "*new ecology*" of design based on a renewed focus on learning from the form and organisation of the models of nature [6, 9].

Alongside the procedural dimension, the aspect of variability represents the second significant principle of design in responsive architecture. This will be clear if we look at the basic aspects structuring and organising design. The code is the principal material used to give voice to this sort of "*living*" organisation. But it is a special code, like DNA in nature, that picks up on the relationship between genotype and phenotype and interprets it in architectural form. Genotype is the basic organisation guiding external input and, at the same time, the "*structural*" place where the input itself gives form to a specific configuration; while phenotype is the result of a physical approach which, starting with the same genotype, may change to take on a final form which is the result of external input from a specific environment. As in nature, therefore, the characteristics shared by these forms of architecture (genotype) constitute a sort of basic structure, the final appearance of

Fig. 1 Soma architecture, One Ocean: Expo 2012 Thematic Pavilion, Yeosu-kinetic media façade consisting of GFRP louvers

Fig. 2 David Fisher, Rotating Tower

which (phenotype) depends on the interactions between the basic structure and a series of parameters which are not resolved individually, which may vary in design to take on multiple configurations consistently with the development of the design. The code dictates the rules of the possible combinations (Fig. 1).

All this is the "*design imprinting*" of digital architecture, as Mario Carpo notes [1].

Looking beyond architecture as the area defined by knowledge, it is a particularity of the responsive approach (and therefore of responsive morphologies) to integrate the dynamic and conformative abilities of an environment in design. In disciplinary terms, the foundations of this line of study are primarily the result of abandoning the dogma of the finite object in favour of an approach to design based on planning and on the diagrammatic dimension of design.

There are two cornerstones in its structure: organisation into basic elements on the basis of a horizontal logic [17], which does not imply a hierarchic sequence of elements in the system, and a set-up consisting of architectural templates demonstrating awareness of the system's behaviour (Fig. 2).

2 Tools

From the methodological point of view, research in responsive design fits into the context of computational design, defined as a design process which takes advantage of the potential of computation, integrating digital and emerging technologies in the process of a product structured on the basis of interaction between form and information. In computational design, therefore, the domains of programming and design are integrated to identify a form of creativity that interprets information in the form of procedures and rules for interaction for the definition of architecture. With the adoption of two synergic approaches: on one hand, a specific focus on the instruments of design and diagrammatic representation, requiring construction of a new system of design primitives of a computational nature, and on the other, interpretation of design as a practice of horizontal interaction which operates in terms of the basic components of design itself.

2.1 Relational Primitives

The methods and instruments of computational design are specifically interpreted in the context of digital thought in their relationship with a number of key elements of its organisation, such as algorithms, parameters and objects, which, due to the recurrent nature of their processes, represent the logical and operative basis of the forms and instruments used in research in the field of responsive architecture [8].

2.1.1 Algorithms

The algorithm is of significance in design because it is an operative procedure based on a form of logic which is independent of the nature of the performer and the language in which it is expressed, in that it lies on a level of abstraction which is higher than the system that will be implementing it. The aspect of algorithmic programming applied to the context of responsive design does not consist so much of the ability to achieve, through guidance, optimal "*control*" of a result (of the expected output of solutions), but of the process contained in the structure of the information in the algorithm, which thus becomes a sort of "*procedure*" for processing the basic components of design, with a capacity for recurrence. This feature determines a common matrix of digital design: its genome [14].

2.1.2 Parameters

Hence the concept of the parameter, the definition of which refers to a datum which modifies a status and its dependants. The importance of parameters in design is demonstrated by the designer's capacity to manage them in order to relate to a plurality of different alternatives on the basis of a single model. This means shifting the focus of design from a single defined object to the process of design itself, which is capable of varying a basic configuration. Moreover, if the concept of the parameter is applied in the context of a scale of values, it means recognising the design object's intrinsic potential to behave in relation to certain requirements defined in the design. These requirements are specifically the product of comparison with a more or less complex series of external conditions.

Context and time are therefore the references within which the game of parameter design is played. Identifying and defining the components and variable aspects of the parameters in a process for the purposes of design underlies the concept of parametrisation, which is defined as the ability to embody the relationship between the fixed elements (limitations) and variable components (parameters) of a single design structure. Having defined a series of limitations on design, parameters may thus freely revolve around a single basic structure in order to generate as many overall effects as there are possible variations in the parameters themselves.

2.1.3 Objects

The definition of the concept of object is more open. In programming, objects are defined as abstract entities destined to endure over time and to be reused in a multitude of different applications in different contexts and environments. This means they are incomplete, differentiating them from the unitary dimension of algorithms, the structure of which is guided by a complete, consistent logic. As in programming, in digital design objects are distinguished by their adaptability, defined as the possibility of continuously moving and organising the environment in which they dwell in different ways. The result is a focus on the dimensions of time and the dynamism of context. In the specific case of architectural design, we must however note that objects are in actual fact to be seen as a metaphor of the concept of the object as defined in the real world, and their material conformation passes through definition of a series of features and behaviours, determining properties and operations which a specific object deriving from its design family may perform (Fig. 3).

2.1.4 Recurrences

Digital design is characterised by a series of recurrent codes and primitive elements of design which make use of algorithms, parameters and objects as the operative tools of their organisation, configured in terms of space in reciprocal interaction. In the geometric definition of design, the result is a series of relational primitives, which in terms of design lead to recurrent instrumental and conceptual actions (if referring to space and to the expressive articulation of the resulting differentiated formal systems) of attraction, discretisation and interpolation.

One fundamental aspect allowing these recurrences to be configured as non-standardised design actions is, once again, their relationship with their context. In the design program, this aspect takes the form of definition of a system of conditions. The term condition defines an expression that may be true or false in relation to a series of assumptions defined in programming, the key feature of which, in relation to a finite sequence of closed passageways, is the opening of the program itself to a system of alternatives. The focus is therefore on the "*behaviour*"

Fig. 3 Dynamic pattern: responsive change in the basic model—Resonant Chamber, Studio RVTR

of the system, rather than on its form. The result is that in assigning this behaviour to the system, the program interprets data and expresses requirements which permit orientation of the direction, force and action of the variables, revealing spatial organisations characterised by an aesthetic of mutation (Fig. 4).

2.1.5 Components

Working with design tools that are not based on the configuration of the single finite object, but on the potential of form that emerges from families of similar objects, is part of the nature of responsive design. Adopting the logic of recurrences, and using mutation as a principle of design, the dynamic capacity of responsive architecture directly undermines the foundations of the solidity incarnated by the stability and unchangeability of construction to reveal a new family of "*intelligent objects*", so-called responsive components capable of interfacing and dialoguing with their context. In design, components therefore become the basic elements of an interactive architecture open to context. These elements are used in the organisation of design on the local level of the individual unit, referring to the overall dimension only as a spatial aggregation emerging from the set of these new intelligent "*bricks*". The relationship between the overall unit and the elements of which it is made is therefore that which leads from an object to a population of similar objects on the broader scale of a structure or entity (Fig. 5). Starting with the simplicity of

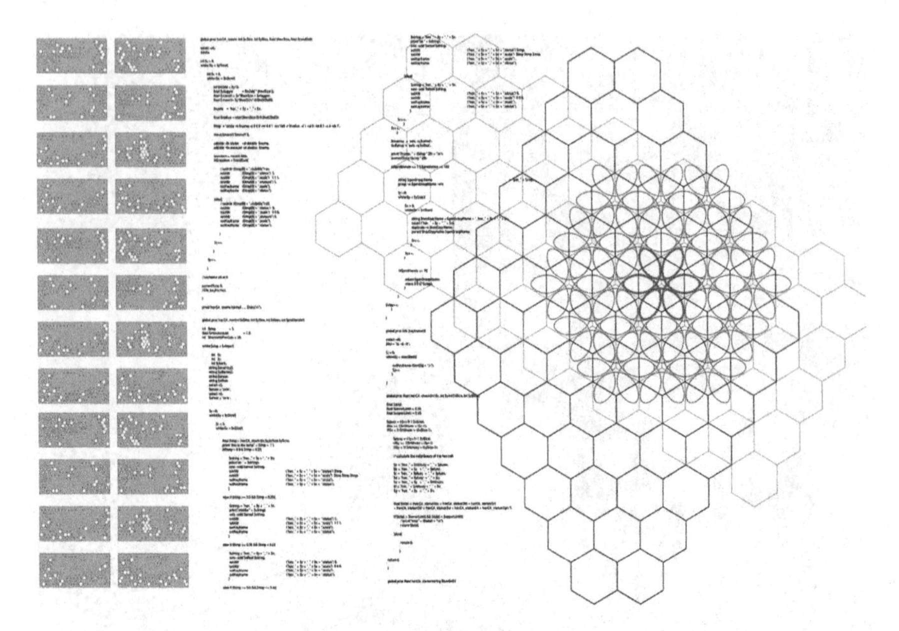

Fig. 4 Responsive pattern: the relationship between basic geometry and the system of conditions (kokkugia)

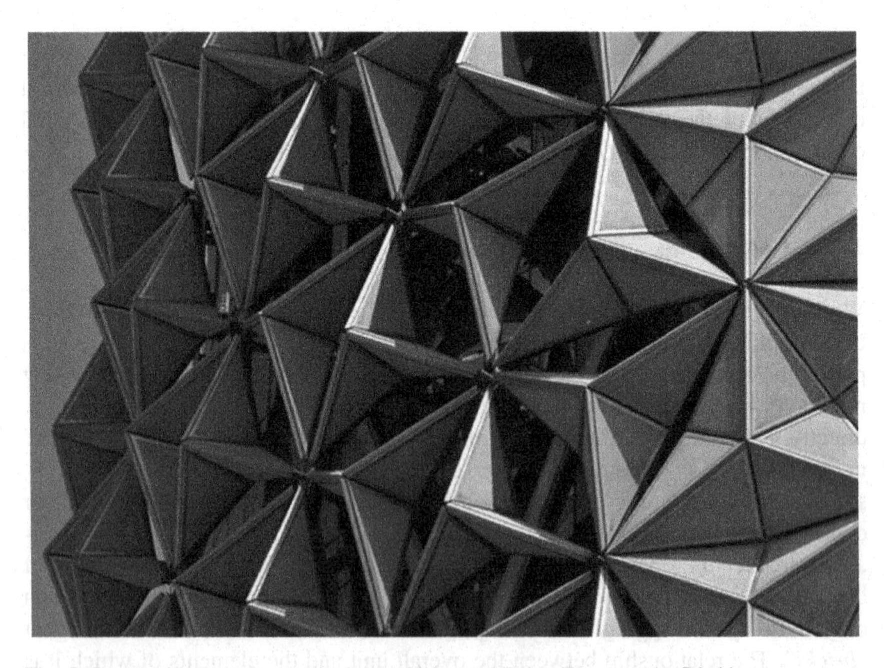

Fig. 5 Responsive components: Aedas Architects, Al Bahar Towers, Abu Dhabi

responsive components, a design is therefore built around the relationship between them, and may design complex, dynamic organisations, some of which may not have been foreseen, in relation to the basic morphology of the specific components of which it is made [5, 12].

3 Experimentation

The considerations raised so far and the theoretical assumptions supporting this study reveal a need to experiment with some of the key elements of responsive design with creation of a prototype of a dynamic surface, as an element of a possible responsive architecture.

Because of the element of novelty which characterises it, in its principles, references and tools, as we have seen in the previous sections, the study of an operative approach to responsive design must therefore be assessed in terms of the procedural potential, reusability of models and operative recurrences that it can generate, rather than in terms of the aesthetic and/or formal aspects of its results.

The resulting applications, definitely far removed from the aesthetic and formal references of the discipline, therefore appear to represent important opportunities to build new methods and tools for study and teaching (Fig. 6).

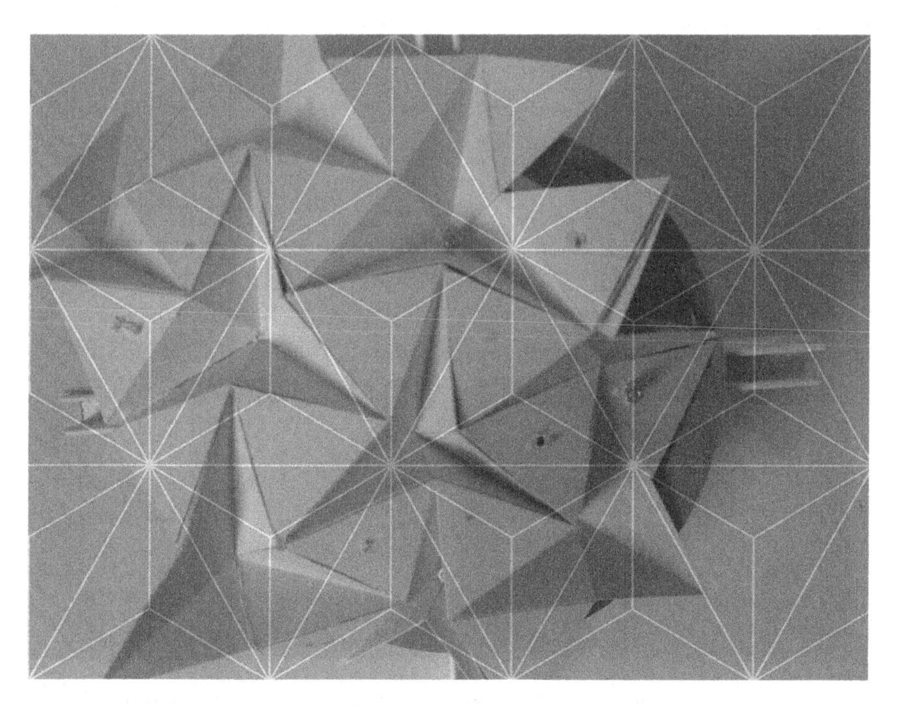

Fig. 6 Origami models for the development of responsive components, diagrams and examples based on a single shared kinematism (Nebuloni, Vignati)

For its aesthetics and logical-constructive organisation, we have chosen a basic origami representing the broader category of folded surfaces to be the subject of study in this experiment (see diagrams in Fig. 7). The experiment focused in particular on *kinetic* (dynamic) *origami*, which may be used to obtain a certain movement by shifting appropriate points (control points). The origami used in the study, characterised by an internal symmetrical kinematic structure, therefore permitted extension and contraction of the basic morphology on a plane and around a common point.

Why origami? First of all, for the intrinsic ability of its form to obtain spatial organisation. And at the same time, for its analogical "*algorithmic*" organisation, made up of rules of composition, and lastly for its ability to achieve a design configuration through simple, coded movements. We may therefore view an origami as a simultaneous blend of rules and creativity. The relationship between architecture and origami, and between design and the practice of origami in general, therefore goes beyond simple aesthetic or stylistic interest in its "*folded*" morphologies. The relationship is deeper than that, based on the common trait of having a structure, general rules and an organisation with which to interact in design.

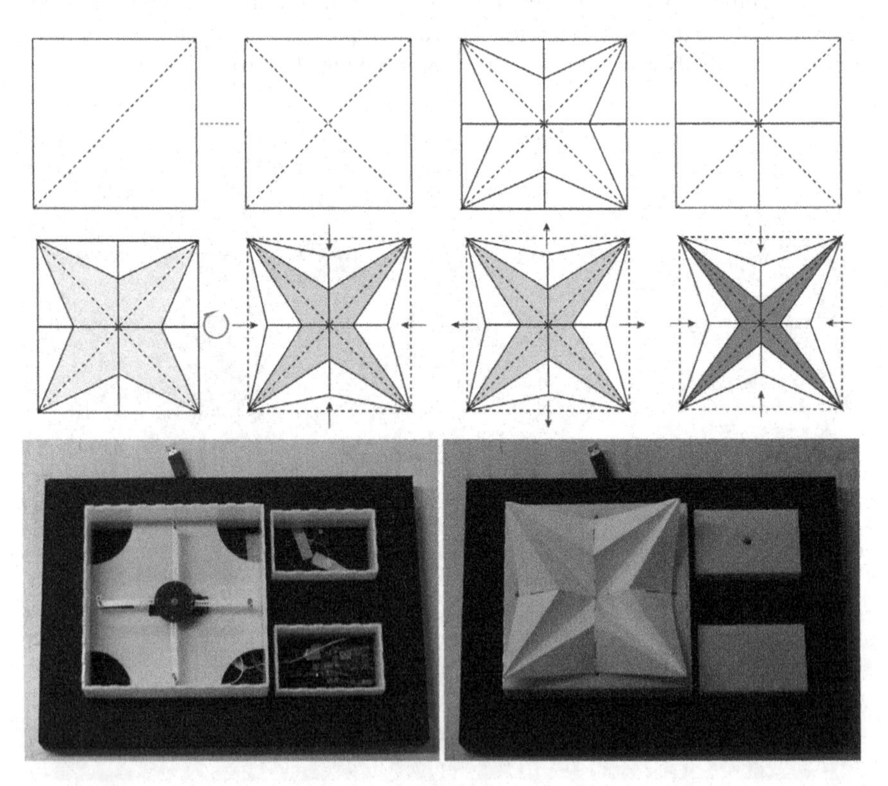

Fig. 7 Diagrams and prototype for the study of a responsive architectural component (from a thesis by Alberto Giacopelli and Antonio La Marca, tutors A. Nebuloni, G. Vignati)

Moreover, like an object in digital programming, origami may be interpreted in view of its broad relationship with a whole family of possible forms, with which it shares the singularity of the internal kinematisms of which it is composed and the particularity of its specific morphology. The essence of origami is therefore a sort of diagram of form and process. Like diagrammatic logic and the processes of algorithmic architecture, origami is therefore characterised by the simplicity of the primary elements of its morphologies and the rules that determine its construction. The possible variation on this in terms of formal organisation always lies within a single basic matrix or family, expressing its morphology through methods and procedures of repetition (once again, recurrences) (Fig. 6).

With regard to its construction, the interest in development of responsive origami lies not in the principles of its composition (and therefore in the "*composition*" of the basic bricks of responsive architecture, or how to "*put togethe*r" the bricks in the final form and set of architecture), but at an earlier level: what object of design can I develop, and how, to produce an open, dynamic architectural component? The interest in origami is therefore not related to the final dimension of the product, but to the programmatic dimension of an object in progress, which may be of great value in terms of architecture and architectural patterns, when the pieces, organised in a composition, encounter a morphology [2].

Before discussing the experiment, let us clarify the meaning of a number of recurrent terms: prototype, in relation to the concept of responsiveness, and Physical Computing, as the disciplinary context within which to consider the instruments and technology of the experiment.

3.1 *Prototype*

For the purposes of the experiment, prototype is defined as a practical tool allowing us to perform tests and assessments of the development of an idea or project. This is definitely an approximate, reductive vision, but it has the goal, and possibly also the benefit, of underlying the practical nature of the prototype. The concept of prototype definitely has a broader, more complex significance, but if we use it we tend to ignore, or even hide, the pragmatic aspects. In other words, construction of a prototype allows us to *reify* and *physically* experience the aspects considered most important in the development of a project, whether it be a product, an architectural or urban manufacture, or an artistic installation. The proposed experiment is intended to use a prototype to reveal the process and the physical elements controlling a kinetic origami surface, and the method and tools used to respond to external stimuli/events. The ability to respond/react to the external environment on the basis of an interaction project further increases the prototype's operative value. In addition to considering form and physical elements (electronic hardware, mechanisms, materials…), it is necessary to design the "*active*" relationship between the prototype and the world outside.

The word *interactive* is often used to underline the relationship with humans. For example, a robot might be considered responsive if it performs its assigned tasks independently (such as those robots which clean the floor on their own), while it is definitely considered interactive if it assists senior citizens, a context requiring communication. In short, it is the involvement or presence of a human component that determines whether a prototype is considered responsive or interactive (Fig. 7).

3.2 Physical Computing (PhC)

The approach on which the experiment is based is supported by the contributions of Physical Computing. PhC is a discipline which aims to construct interactive systems through use of hardware and software. Hardware is defined as a set of electronic and mechanical components, while software is a series of programs developed to control the behaviour of the system.

In this context, the system reacts by using the physical properties of the world (mechanical, chemical, electrical… and through devices such as sensors and actuators) and the ability to process data which determines how a system reacts. This is obtained by codifying behaviour in a programming language.

In other words, the responsive system receives a stimulus (captures/detects physical properties in its environment as input), prepares a response as output and implements it by modifying the physical properties of the external environment using actuators, devices which "*actuate*" the response (Fig. 8).

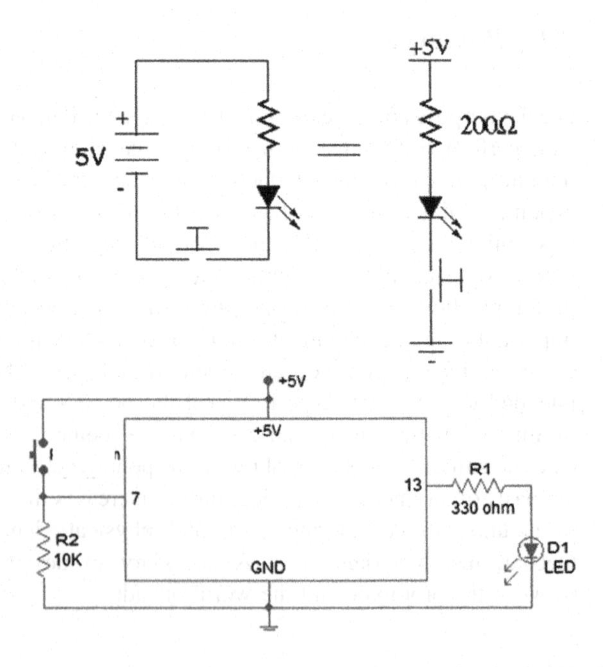

Fig. 8 Wiring versus computing circuit

This is a crucial point. Figure 8 represents the two possible ways in which a system may respond to an external event. The aim of the system is to turn on a light source (represented by a LED) with a button that closes the circuit if pressed and opens it if released. When you hold the button down (the system detects external input), current circulates in the circuit, and the LED comes on and stays on until you release the button. The first system achieves this goal using a wired circuit, and the only possible behaviour is LED on for the time in which the circuit is closed, or LED off for the time in which the circuit is open. No other functions are possible. If the specifications change, for example, if you need a LED with a luminosity transition or flashing at a variable frequency, you must change the circuit and add other components, that is, change the hardware. The second system can perform a number of different actions, with appropriate code in the processing component: read the status of the button, calculate closing or opening time, count the number of openings and closings, and more. The system thus configured will cause the LED to behave in a way responding to the input data, such as flashing or changing its luminosity.

Once malfunctioning of the input and output components has been excluded, the distinctive feature of a responsive system, in the context of the PhC, is physical separation of input and output and connection of the two through processing. Flexibility, modularity and extensibility are features associated with the use of code, so that different functions and behaviours can be achieved without changing the configuration of the physical system [11].

Much of the merit for the popularity of Physical Computing lies with the context in which it has developed in recent years. Essential factors such as the miniaturisation of electronic devices, the availability of software development environments, lower costs and the increase in processing power have contributed to its popularity. Another key has been use of the web as a catalyst, permitting the birth of numerous very vocal open communities and the resulting distribution and sharing of theoretical and above all practical knowledge, in which projects and experiences are discussed and made available in open form, or at least free form, permitting active participation of a large number of players, engineers, architects, designers, artists and makers [16].

Through the communities present on the internet, it is possible and even advisable, when developing an idea or project, to start by exploring similar solutions from which to freely draw ideas in order to experiment with and replicate their experience, studying it and understanding it in order to come up with a solution of your own which improves on or expands the original one. This approach applies to both software and hardware components. The key principle is that it is no longer necessary to "*reinvent the wheel*" [7].

3.3 Experimentation: Redesign

In addition to the didactic research in architecture promoted and developed by the Physical Computing laboratory (Phyco.Lab) at Politecnico di Milano—Department of Design, experimentation in this area has seen an initial phase of collection and study of case studies, resulting in identification of a number of significant experiences that achieve the desired goal. After this, their components were studied, broken down and analysed, focusing on the movement system, and on the basis of this, goals were reconstructed and the kinematic component was redesigned. We opted for construction of a modular, flexible basic environment which is easy to build, economical and quick to assemble. Responsiveness was obtained using the Arduino platform and re-processing samples of code in a template-based approach (Figs. 9, 10).

The original examples form a simple responsive system with use of origami as a formal element of the morphological component and simple input and output sensors. The first prototype (Fig. 9) presents a responsive origami controlled by a proximity sensor. The result is highly effective despite its simplicity.

The second prototype (Fig. 10) regards a plan for an interactive wall composed of panels made with origami elements; in this case the sensor used is a light sensor. Note that the two examples share similar movement mechanisms.

These models/templates turn a servomotor with an angle expressed in degrees, the value of which depends on the amount of light in the room, in one case, and the proximity of an object (such as the user's hand) in the other. The functional structure the two share is summed up in the figure below, where, without losing its

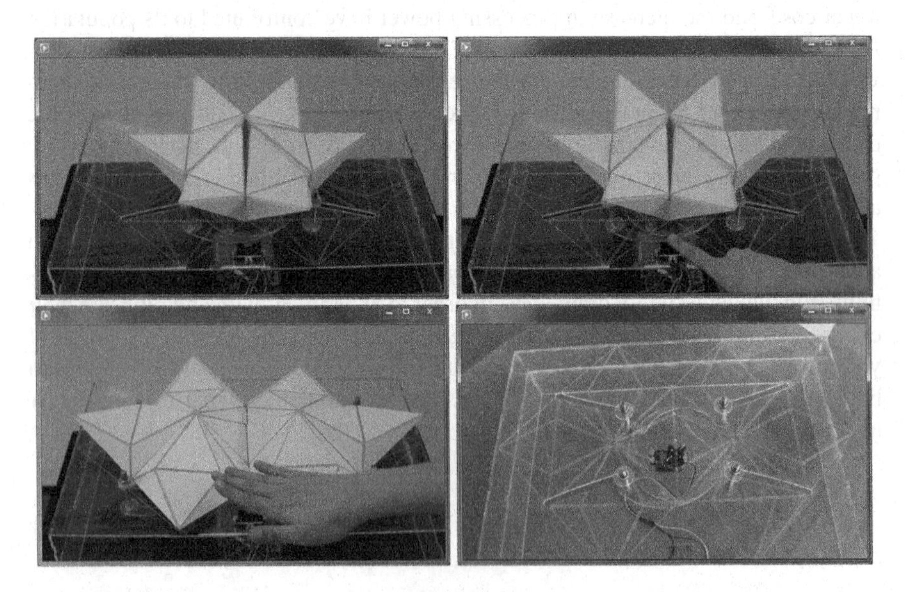

Fig. 9 Prototype of origami responsive to proximity

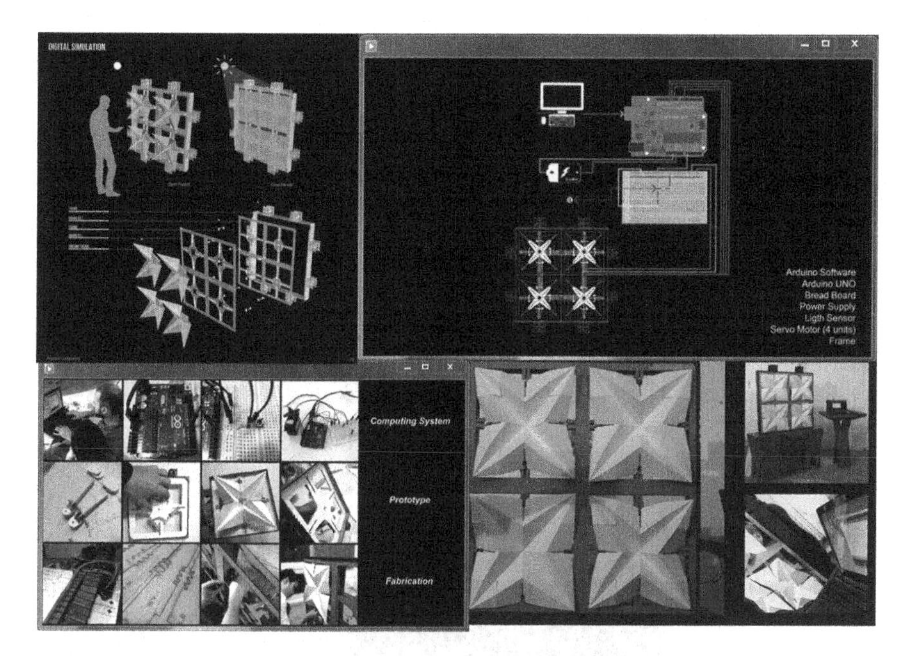

Fig. 10 Responsive wall: concept, prototype scheme, development process, installation

general nature, a photoresistor or LDR (Light Dependant Resistor) is used as a sensor (Fig. 11).

The circuit scheme is shown on the left, where the light sensor symbol may be seen, along with the controller (Arduino) and the servomotor symbol. In the middle we see the development environment (IDE-Interactive Development Environment) and the code implementing the system's behaviour; on the right is the physical prototype, containing the real components (sensor, Arduino, servomotor). Without going into the details of the code, the two instructions codifying its behaviour are shown here: (analogRead ()) which reads the light sensor and converts it into a value (val), which is then appropriately transformed into degrees and passed on to the instruction (servo.Write ()), which controls servomotor rotation in degrees.

As we may see, the prototype does not take on any particular form here, and its minimalism expresses the system's functions (behaviour).

The system thus configured has an intrinsic modularity of its own which allows the light sensor to be replaced with another type of sensor (proximity, humidity, distance…), with only minimal changes to the structure of the code, other than the specific functions involved in reading the sensor. Once the responsive component of the experiment has been identified, we shifted our attention to the components (mechanical/kinetic) that physically implement the movement. The study was conducted in the following phases:

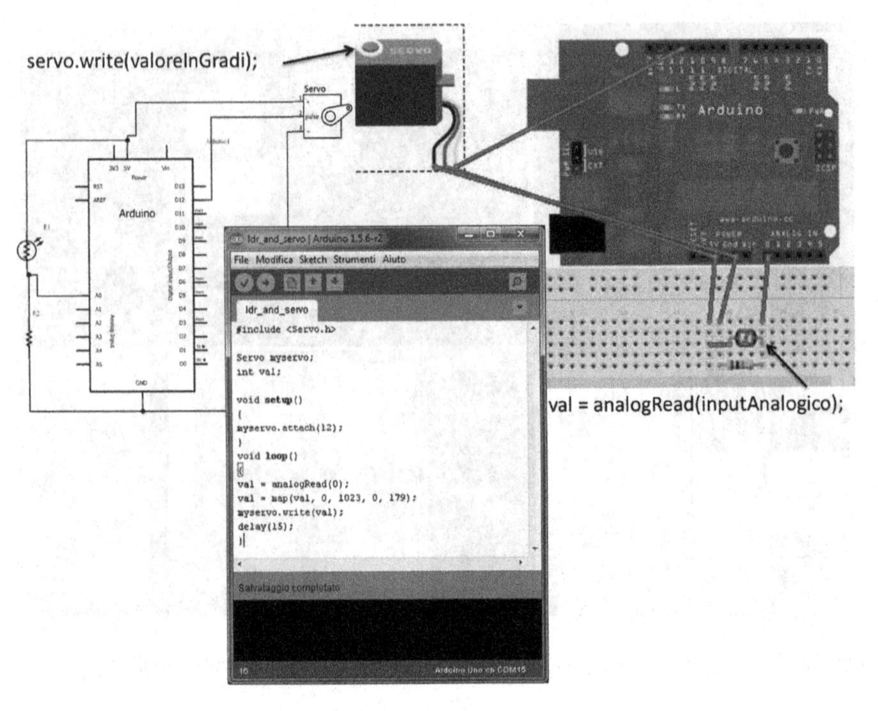

Fig. 11 Structure of a responsive application

1. analysis of the mechanism used in the projects referred to;
2. modification of the mechanism and motivation of the choice;
3. geometric simulation of the mechanism;
4. geometric simulation of origami;
5. construction and assembly of components.

Before going into the components, a brief introduction to the principal terms characterising the elements in motion is required: mechanism, kinematism (or kinematic chain) and joint. Here is a brief definition:

- Kinematic chain: it is a linkage of elements and joints that transmit a controlled output motion related to a given input motion;
- Mechanism: it is a kinematic chain where one element (or more) is fixed to the reference framework (which can be in motion);
- Joint: guarantees the contact between two members and constrains their relative motion.

The term kinematic chain is sometimes used to refer specifically to a set of elements in which, having established the position of one element, the position of all the others is determined. There is therefore a univocal relationship between the trajectories of the various parts. If one element in the kinematic chain is immobile (acts as a frame), it is normally referred to as a mechanism.

3.3.1 Analysis of the mechanism

Abstracting the above definitions in a geometric context, we may consider the mechanism as a physical system (device) which transforms a main curve into a derivative curve. The important thing is the law of transformation which describes how the movement of a point on the input curve is translated in the output curve. For the purposes of the prototype, the analysis was applied to both the geometric elements of the mechanism and assessment of its physical implementation. The diagram of the mechanism implemented in the examples consists of a connecting rod kinematism (Fig. 13). The function of transfer between angular rotation and shifting of the connecting rod is not linear, and the latter depends on the cosine of the angle of rotation.

The mechanism constructed assembles a number of components (connecting rod, cams, guide and slide) through two joints and a supporting frame. We may suppose that the final system will have a relatively long assembly time, as well as a certain rigidity in modification and maintenance (Figs. 12, 13).

The conclusion is that due to the material used and the number of components, the connecting rod mechanism is not appropriate as an easy environment for experimentation which can be assembled quickly and is flexible (for instance, for changing the mechanism).

Fig. 12 Responsive origami: interaction with proximity sensor. Mechanism that moves the origami and interpretation of the mechanism diagram

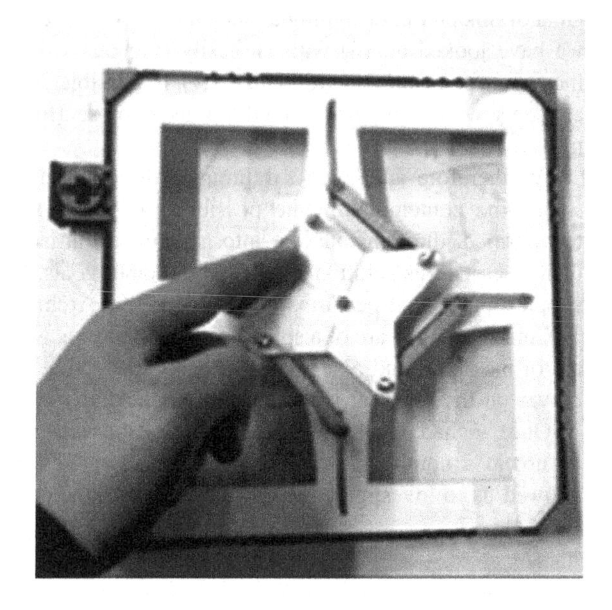

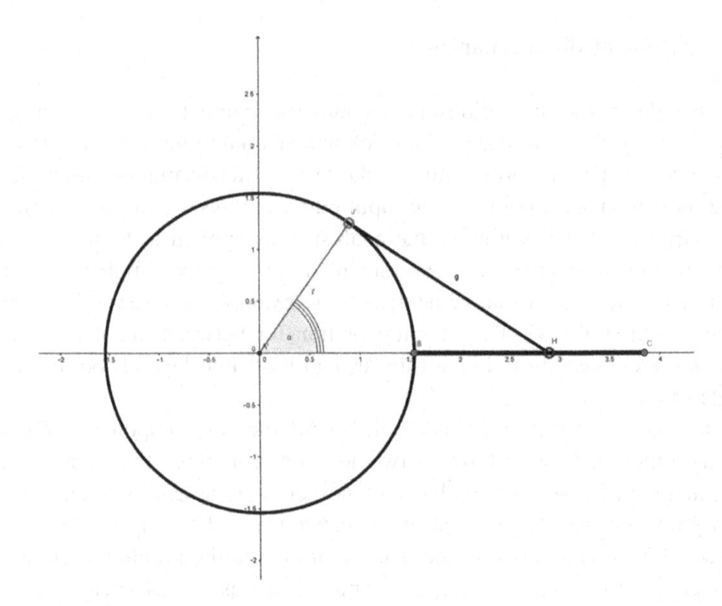

Fig. 13 Diagram illustrating the kinematism

3.3.2 Modification of the Mechanism

Goal of this phase: designing a mechanism (i) with less elements than the examples we have looked at; (ii) with an easily changeable input curve; (iii) made out of inexpensive material (cardboard); (iv) if possible, made out of easily available elements which can be adapted for its purpose (for the slide and guide in the model).

We therefore attempted to eliminate the connecting rod element while at the same time contemplating the possibility of using an input curve that could be transformed, through rotation, into a segment (an element of a straight line). We then assessed this solution in view of the other goals. In geometry, there is a single simplest curve that performs this function, the Archimedes spiral (Fig. 14).

Considering the arc of a spiral as the input curve, we need only two elements, the arc of the spiral and a segment, with no connecting joints. The only element that moves on the segment is the element connecting the two curves.

Once we had determined the curve, we sought responses in mechanics, verifying its use in a number of different mechanisms [13].[1] The next phase in the study allowed us to investigate the change and draw the arc of the spiral.

[1]Cfr.: Brown H.T (1868), Five Hundred and Seven Mechanical Movements, Virtue & Yorston-Hiscox G.D. (2000-ristampa), 1800 Mechanical Movements, Devices and Appliances, Lee Valley Tools. http://engineering.myindialist.com/2011/kinematics-of-machines-tutorials-classification-of-cams-and-followers/#.VjeVA2tsxZQ.

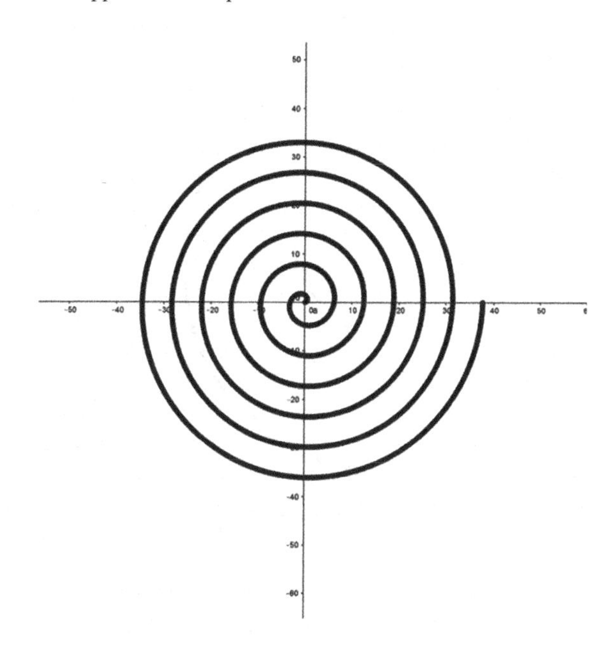

Fig. 14 Archimedes spiral

3.3.3 Geometric simulation of the Mechanism

Before physically constructing the mechanism/kinematism, we conducted a simulation using GEOGEBRA (GEOmetry and alGEBRA) open source software, which, among its many features, allows us to build (and animate) relationships between geometry and algebra, both on a plane and in space. Use of this tool (which is not a modelling CAD, but an interactive environment for simple, effective exploration of the concepts of geometry and algebra), makes it possible to build and analyse geometric relationships between the curves defining the mechanism. Moreover, it is possible to simulate the movement described from the points on the curves which transform motion. Working on construction of a prototype, rather than mechanical simulation, our interest focused on the relationships between the elements that convey motion. We therefore constructed a diagram of the mechanism, setting aside physical properties such as mass, friction and the forces in play for the time being. The approach was to use simple tools to determine the geometric correctness of the proposed mechanism, leaving choice of the physical components and functional tests to the final phase of construction.

Figure 15 show the geometric elements underlying the mechanism: a spiral, a segment and the point at which they intersect. During construction these elements will correspond to physical elements, as described below. The spiral is the input curve, the segment the output curve, and the point at which they intersect transforms the spiral into the segment. In greater detail, the spiral is connected with the

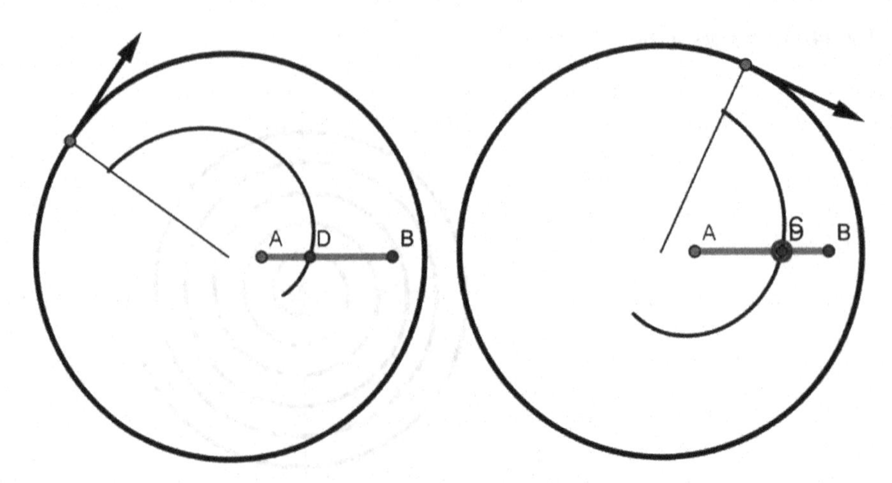

Fig. 15 Diagram of spiral kinematism used in the prototype studied, with a single arc

segment through a connection geometrically represented by the point at which they intersect. Rotation of the spiral has the effect of changing its intersection with the segment. In other words, rotation of the spiral curve moves the point on the segment, creating the mechanism. The relationship between the angle and the shift in the point is described by the geometry of the spiral: the distance travelled by the point is proportionate to the angle of rotation.

3.3.4 Geometric Simulation of Origami

The scientific literature includes a large number of studies of origami, most of which are concerned with algebra, geometry and algorithm theory.

As we have seen, Geogebra is not a modelling tool, but it has the ability to construct and display parametric geometries/curves and the geometric relationships between them. In the case under examination here, this potential is used to virtually and parametrically assemble a very simple origami (Fig. 16).

By appropriately building relationships between auxiliary geometric elements (circumferences, straight lines, intersections, right angles…), we created the basic element in the origami (in this case, a triangle), which we than assembled in its complete configuration using appropriate symmetries. In this way rotation of the arc of the spiral moves the point along the segment, and the segment moves the point of the triangle which creates (simulates) the kinematism intrinsic to the origami itself.

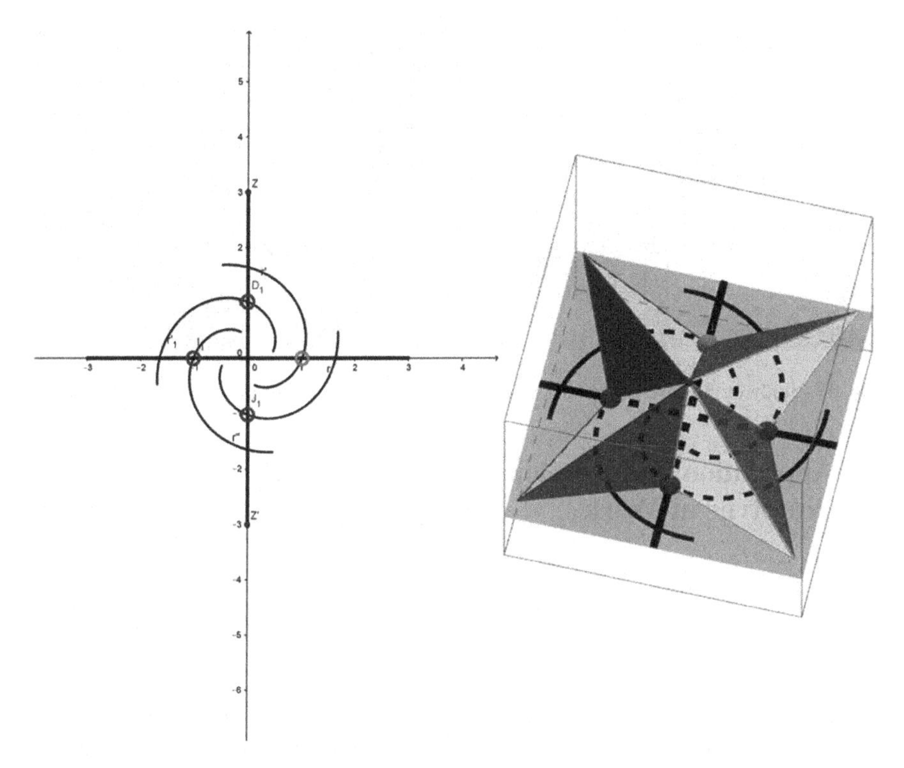

Fig. 16 Simulation of kinematism and origami

3.3.5 Construction and assembly of Components

Having completed the simulation phases, we then approached construction of the mechanism, seeking to use inexpensive, readily available materials. The electronic components (microcontroller, sensor and motor) maintained the configuration of the examples referred to, that is, an LDR sensor as the element varying its resistance on the basis of the level of light in the room and a servomotor capable of turning at an angle of 0–180 degrees. The spiral arc guide was made of cardboard, while the slide and the guide were created by coming up with an alternative use of electrical ducts, also used to fix the kinematism to the servo pin. We used two-sided tape to anchor the ducts to the frame, which was also made of cardboard (Fig. 17).

At this stage we checked the practicality of our decisions regarding the form of the spiral art, finding a compromise between weight, cardboard-pin friction, servo power, and the size and movement of the origami. Assembly of the prototype took about twenty minutes.

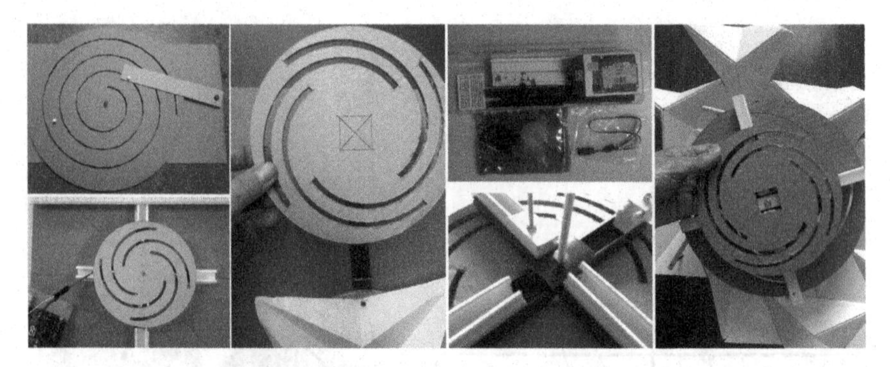

Fig. 17 Study of the basic kinematism, development and assembly of the model and alternatives

4 Application and Development of the Model: Prototype at Work

What I cannot create, I do not understand

(Feynman R.)

The phase following construction of the basic prototype led in two different directions, representing integrated aspects of the research: one which is educational and methodological, and another concerned with design and application.

The educational aspect specifically regarded the areas of communication and comprehension of the technical and instrumental part of programming (platforms, algorithmic interfaces and modelling, coding languages and techniques).

Interest in the project focused on the capacity for processing and development of complex morphologies on the basis of a logic of population and variation of the overall dimension of a basic responsive element.

The result may be measured not so much in terms of project's ability to offer a return on the basis of the canons of representation as in terms of the efficacy (and propensity) of its variables for simulating behaviour and forming a system.

The experimentation environment that had been built was then formalised in a teaching kit used to teach the rudiments of responsive technologies and to study other interactions and surfaces (origami or other types).

The kit was used in various workshops in the "*responsive morphologies*"[2] series and during the 3rd International-Regional workshop eCAADE[3] "*Computational Morphologies*" (Figs. 18e, 19).

[2]Courses in the Design Department at Politecnico di Milano (under the authors' direction and coordination) in the teaching program of the Physical Computing laboratory (Phyco.LAB).

[3]Education and research in Computer Aided Architectural Design in Europe. 3rd eCAADE International-Regional workshop, Milan, May 2015.

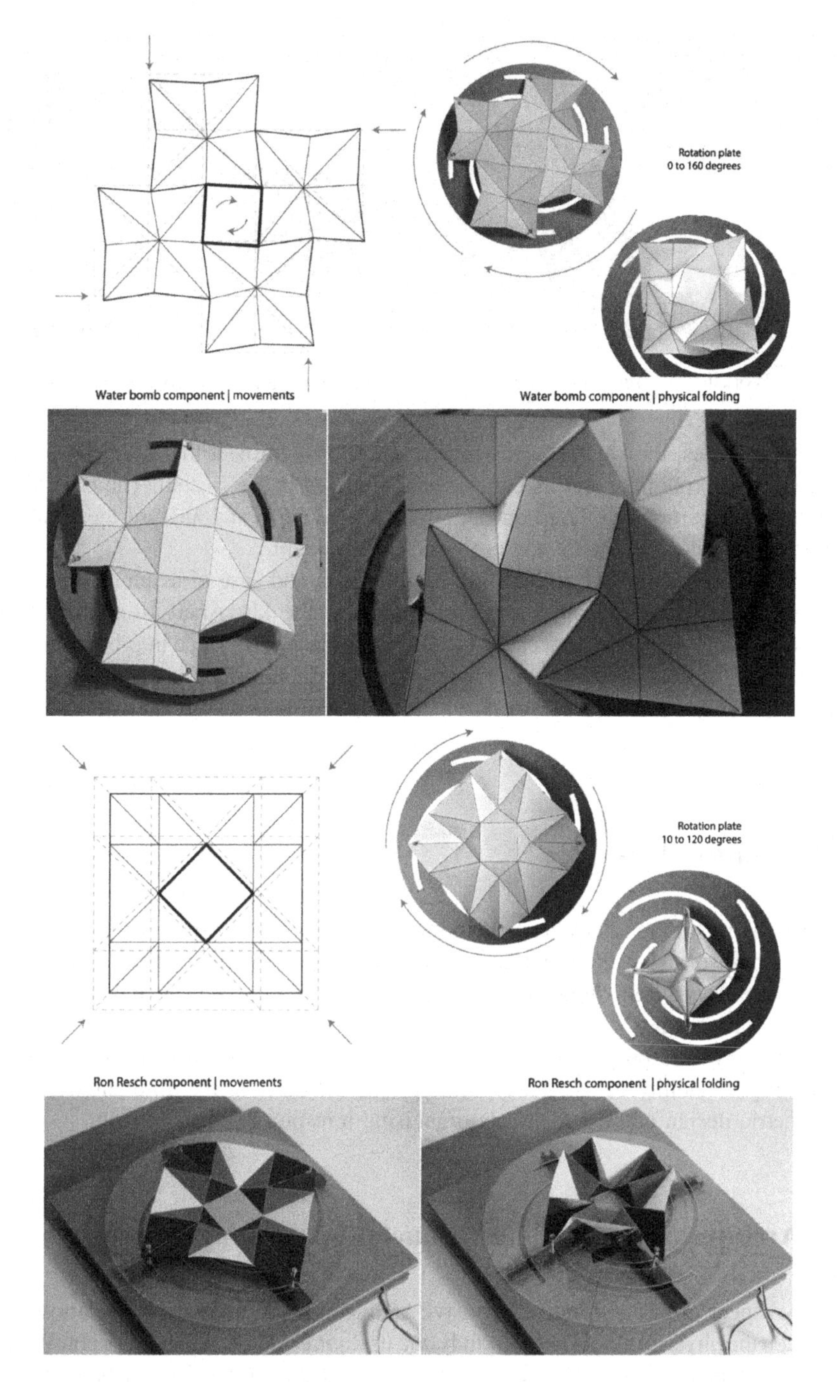

Fig. 18 Prototypes of responsive components developed from a single basic kinematism (project by Andrea Quartara, from workshop in "responsive morphologies"—Nebuloni, Vignati—PhycoLab)

Fig. 19 Workshop in "responsive morphologies" (Nebuloni, Vignati—PhycoLab)

In both cases, the key goal was to introduce to the project the basic aspects of responsive morphology and its components involved in the system. The eCAADE workshop in particular offered an opportunity to identify and analyse the positive aspects and critical points of the proposed environment.

The positive aspects were certainly its simplicity and ease of assembly, and the efficacy of the template approach proposed with the kit. Apart from the small amount of time available, especially for those with less experience in programming, the critical aspects were use of low power servomotors and anchorage of the kinematism to the motor (an aspect which was modified in implementation of the model).

4.1 Future Work

Directions for further study in the experiment include the possibility of modifying the kinematism using the input curve and variation of the number of control elements increasing both input and output curves (for example, by developing an origami morphology with five control points—Fig. 20).

The study thus proposes development of the results obtained with the goal of achieving scalability of the study prototype, on the basis of the different scales of parametric design and the materials organising a responsive environment.

5 Architectural Templates: Conclusion

A key aspect of research into responsive architectures is the renewed relationship between the physical model used in study and its digital interpretation in design. While in one case the focus is on the material aspects and the overall relationship between the components in play in the design program, in the other case the aspect requiring attention is implementation of parametric programming of the design itself.

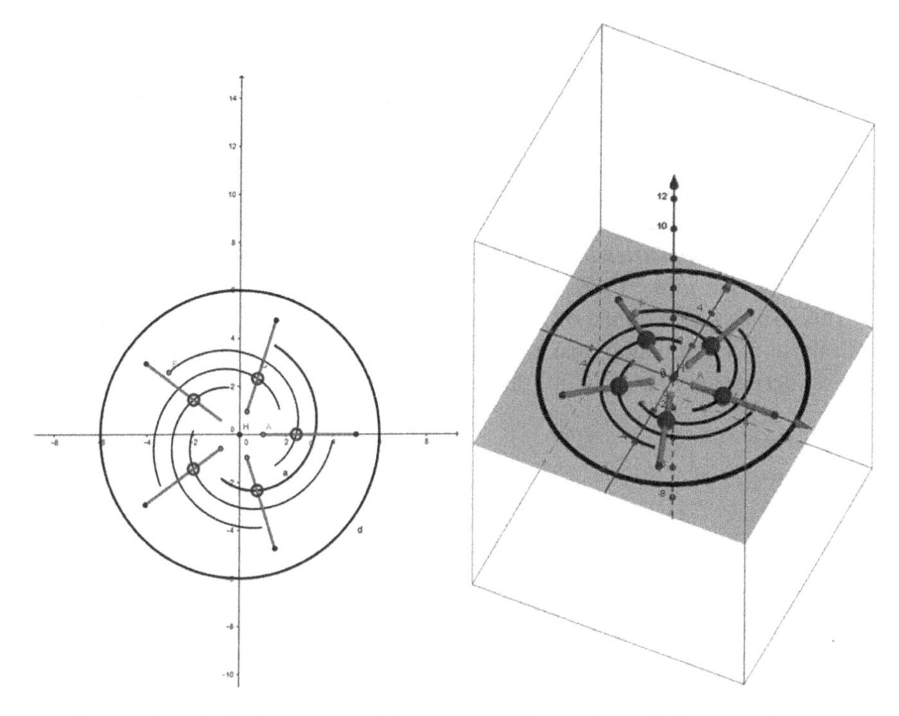

Fig. 20 Variations in the kinematism of the prototype: kinematism with five control points

Rather than instrumental, the challenge of design is therefore methodological in nature, finding no useful, immediately replicable references in the consolidated context of the discipline.

This implies the simultaneous presence of moments and phases in development of the design which are not homogeneous, which are normally addressed separately on the basis of an incremental, linear study logic.

The idea therefore becomes architecture not through application of a structured preliminary process of critical analysis and interpretation, but through on-going experimentation with objects, in which the expressive diversity of their languages reveals alternative ways of seeing and interpreting the design. And in all this the conceptual model appears to be the tool that permits the greatest inclusion and maintenance of those aspects of variability which revolve around the initial idea and take form in the parameters of its generation.

Simultaneous adoption of a digital model and a physical model thus shifts the focus of design into a context which has not been fully defined, in which the objective is not the ability to represent architecture but the ability to simulate its behaviour in general. Whether it be a matter of prototypes or simply models for the study of the overall organisation of space, these responsive "*objects*" are not intended to unify and harmonise the various parts of the composition in a single form, but remain separate elements taking part in the project. Just as in the logic of templates, models of responsive design components are halfway between a diagram and a formal definition of architecture, revealing certain parts of its organisation and how these conform in terms of space, but without acting as closed, defined objects.

The specific feature of the relationship between analogical and digital also permits incorporation of the physicality of design directly in the design process and, at the same time, takes it onto a more directly communicable level which "*makes less mysterious the transition between intent and results, between interpretation of reality and the geometries asked to transform it, between figures and instruments for manipulating them*" [3].

This means that the time of return and the aspects of simulation are bound up with those of construction, making analogical preparation of the study and reflection on design the first space for the feasibility of digital design.

References

1. Carpo, M. (2011). *The alphabet and the algorithm*. Cambridge: MIT Press.
2. Casale, A., & Valenti, G. (2012). *Architettura delle superficie piegate. Le geometrie che muovono gli origami*. Roma: Kappa Edizioni.
3. Corbellini, G. (2013). Ultimo tango a Zagarol. In *Corbellini G. e Morassi C., Parametrico nostrano, Lettera ventidue, Siracusa.*
4. Fox, M., & Kemp, M. (2009). *Interactive architecture*. New York: Princeton Architectural Press.
5. Hanna, S., & Schleck, A. (2007). *Embedded, embodied, adaptive. Architecture + computation*. London: UCL Emergent Architecture Press.
6. Kolarevis, B., & Parlac, V. (2015). *Building dynamics: Exploring architecture of change*. London, New York: Routledge.
7. Igoe, T. (2011). *Making things talk*. Beijing: O'Reilly Vlg.
8. Nebuloni, A. (2015). *Il progetto imperfetto. Lo studio della forma nei modelli di organizzazione generale dello spazio*. Santarcangelo di Romagna: Maggioli.
9. Nebuloni, A. (2014). Progettare architetture responsive. In *Uno (nessuno) centomila, prototipi in movimento. Trasformazioni dinamiche del disegno e nuove tecnologie per il design, Rossi M. e Casale A., Maggioli, Santarcangelo di Romagna.*
10. Oosterhuis, K. (2012). *Hyperbody. First decade of interactive architecture*. Delft: Jap Sam Books.
11. O'Sullivan, T., & Igoe, D. (2004). *Physical computing: Sensing and controlling the physical world with computers*. Boston: Thomson Course Technology PTR.
12. Reas, C., & McWiliams, C. (2010). *Form + code in design, art, and architecture*. New York: Princeton Architectural Press.
13. Roberts, D. (2000). *Making things move: DIY mechanisms for inventors, hobbyists, and artists*. New York: Tab Books.
14. Terzidis, K. (2006). *Algorithmic architecture*. Oxford: Architectural Press.
15. Trebbi, J. (2013). *The art of folding: Creative forms in design and architecture*. Barcelona: Promopress.
16. Vignati, G. (2014). Prototipi responsivi. L'approccio nel contesto del Physical Computing. In *Rossi M. e Casale A., op. cit.*
17. Wiscombe, T. (2014). Discreteness, or towards a flat ontology of architecture. In *Project—A Journal for Architecture, 3/2014, Consolidated Urbanism, New York.*

Architecture After the Digital Turn: Digital Fabrication Beyond the Computational Thought

Andrea Quartara

Abstract Since the Nineties architectural thought employs the new language of software to describe, design, predict, simulate and evaluate form. Lately, digital technology has been enhancing new productive equipment: the virtually designed forms become prototypes and physical models. Computational design bequeaths a post-digital need for material and for fabrication: it's the File-to-Factory age. Digital fabrication technologies actually have a deep impact on the fulfilment of architecture and on its design methods, above all. These facilities increase the level of control that architects have over the designed and consequently materialized architectural form. "*Digital materiality*" (Digital materiality in architecture, Zürich: Lars MüllerPublishing, 2008) turns the digital into physical. This story telling uses some explanatory design experiences relating to the challenge of the traditional thinking of design. It explores the real changes in the way architects design and build the physical environment and foresee how to bring them into the real architectural practice.

Keywords Digital turn · Computation · Performance · Building material · Robotic fabrication

A. Quartara (✉)
Department of Architecture and Design, Università di Genova, Genoa, Italy

M. Rossi and G. Buratti (eds.), *Computational Morphologies*,
https://doi.org/10.1007/978-3-319-60919-5_9

1 Defining a Background for the Digital Fabrication in Architecture

Over the past few decades, computer and digital environment have broadened architecture as a discipline: PC and software became the channels through which the research addresses a Copernican revolution. Information Technology (IT) media rapidly evolves and increases its computing power in very few years. Born as the digital evolution of the drafting table, new digital tools turn into the engine of an emergent design method change.

The main interest of this essay is facing the post-digital need for materiality in architecture. Arising from the everlasting relationship between architecture and geometry, this essay underlines the meaning of geometry as an eidetic multidisciplinary image and, above all, as an ordered system for the digital fabrication of architecture. "*Architecture's desire for geometry is for substantiation by both a material and sensory discourse, and at the same time by an abstract and cognitive formation*" [1]. Davidson and Bates [1] expressed in their editorial of "*Architecture after geometry*" that to be after geometry means to acquire the initiating strategies of maths and logics that the digital technologies (both software and hardware) can expand. Since the 80s, architecture deals ever more with data and information: manifold information, involved in the early phase of design, drive the form finding process. Information is matter of the first digital turn in architecture and its astonishing spread makes of it the new fifth dimension of digital architecture.[1] C. Alexander in his book "*Notes on the Synthesis of Form*" (1964) underlines the need for rationality in architecture: "*To match the growing complexity of problems, there is a growing body of information and specialist experience. This information is hard to handle*" [2] and it needs to be organized. The diagram helps the designer to overcome the difficulties of the complexity because it is a graphical representation that classifies different aspects. Diagrams systematize the design workflow: they serve as a picture of the designer's view of some specific informed problem aiming

[1] S. Giedion with his book *Space, Time and Architecture* (1941) added time to the three dimensions of space (both real and virtual): these four aspects with information are the founding features of digital architecture.

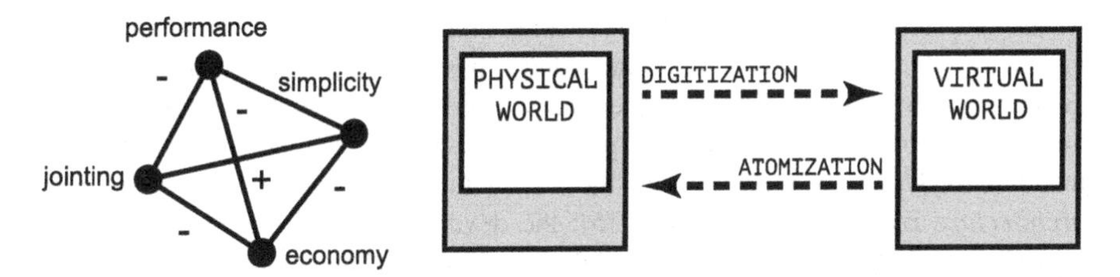

Fig. 1 A typical design diagram by C. Alexander: showing the connections between general requirements of a design problem he shows the need for rationality achieved thanks to diagram (*left*). The interface between virtual and physical worlds: the complimentary yet opposing processes of digitization and atomization underlines the meaning of being digital in architecture (*right*). Author's reproductions

at the "*goodness of fit*" [2]. This kind of programmatic design (Fig. 1) is enriched by digital tools, but the real functionality of a computer does not lie in its hardware powerfulness rather it is defined by the designer's abstraction. The architect takes advantage of the software logic in order to extend human capability. The design approach that demiurgically produces several drawings like plans and sections is powerfully overcome by the digital algorithmic logic, which has in diagram its archetypal matrix. Sets of information, from manifold sources and of different nature, are combined within the coded workflow: the digital actuation of algorithm is the pivotal element of the architecture's language overturning. It is a highly formal language; indeed it interprets the information based solely on its built-in rules. However the computational feature radically expand the power of the diagrammatic process. The designer has a twofold role: first of all developing the initial logics for the generative processes and secondly evaluating the outcomes. This improvement results from "*the programmatic clarity in the designer's mind*" [2]: the algorithmic approach does not produce homogeneity, but rather a balanced world that "*tries to compensate for its own irregularities by fitting itself to them, and thereby takes on form*" [2]. When efficiently improved in a digital non-linear design process, the logic of diagram makes advantage in the interconnection of different disciplines, such as material science, biology, and engineering till the physical fabrication. Digital forms overcome the combinatory feature of "*diagram of forces*" [3]: the architect exploits the software language to shape rules reaching unexpected informed forms that meet the requirements of the input parameters. Kwinter (2003) spoke about "*an invisible matrix, a set of instructions, that underlies—and most importantly, organizes—the expression of features in any material construct.*" [4].

In this territory between chances, determinacy and potentiality the digital technology allows designers rethinking what architecture can express. Digital tools, both hardware and software, enlarge the procedural role of diagram: IT moves away from being a mere tool of geometric representation, reaching the identity of digital design method. From this, during the 90s, the digital culture yields architectural works and researches after[2] geometry. The aforementioned "*Architecture after geometry*" [1] AD magazine issue provided a basis for the reaction of the subsequent years. More than a decade after the seminal AD editorial, Meredith revisits the turn of expression in his article "*After After Geometry*" [5]: he employs this pun in order to depict architecture after the digital turn. "*At some rudimentary level, architecture remains about matter*" [5]: the developments of digital tools trigger remarkable shifts not only in virtual design, but also, and above all, in digital fabrication and manufacturing of forms.

Software technologies facilitate automated processes and new solutions for complex—not complicated—forms. However, "*the most powerful and challenging use of the computer [...] is in learning how to make a simple organization [...] model what is intrinsic about a more complex, infinitely entailed organization (the natural or real system)*" [6]. So the "*real revolution is about the change of thinking, not just the technology improvement*" [7]. As a tool of analysis, computation can provide insight to better understand the nature of something that already exists. On the other hand as a tool of design, it can bridge different worlds, from biology to engineering erasing the break up between design and material realization. Computational models can incorporate simulation, optimization[3] and evaluation already since the early design phases: algorithmic code allows achieving scientific and more and more optimized results—but they are always local, reductive and incomplete solutions. That is, the focus of interest is not the enhanced virtual bits [2], neither the process nor the shape itself. After an early enthusiastic evolution and a subsequent fertile spread, S. Kwinter observes the "*computational fallacy*"[4] [6]. A thumbnail timeline of the computational growth underlines a shift from the early form-based design to the performance-based design. Under the computational point of view, architectural form simply acts as a deployed logic. Under the remarks of the Canadian-born architectural theorist, the research switches from the performance-based design to a material-based design. Digital practices born in the early 1980s as a new way to represent, undertake the role to connect digital architectural designs to their materialization (Fig. 1), to turn "*bits into atoms*" [8]. The historical relationship between architecture's representation and its realization re-emerges from the relentless virtual abstraction. The advanced computing practices are moving away from a representational virtual mode, focusing on the

[2]"After" is used following the meaning previously suggested by Davidson and Bates [1].

[3]Computational tools implement Genetic Algorithms (GA) to mathematically translate the natural rules of evolution; therefore architects can improve design with an open-ended form finding. The evolution process directly involves many cornerstone of biology, as fitness, genotype, and phenotype, own of performance-based design.

[4]S. Kwinter first published this statement in *Thresholds* (2003). He expresses a deep disjunction between the use of computerized methods in design and the computation. Therefore, the theoretician considers the increasing pervasiveness of computation in architecture. As a mechanistic process, Kwinter explains, the numerical processes become over-abstractions of physical processes. The consequence is an imposed order whose reality can be realized only through means of translating the representational to the physical.

physical production. "*Paradoxically, with pervasive digitalization the materialization of the digital becomes the focal point*" [9]. Meredith [5] recalls the "*computational fallacy*" stated by Kwinter (2003): digital in architecture is not limited to a field of language or representation. Digital methods require a material instance: this need finds the fertile ground of growth in the updated digital making tools. After years of 3-dimensional digital models and evolutionary virtual optimizations, the interest in physical form re-emerges. We are witnessing the post-digital need of architecture for matter: it finds expression in the wild fire diffusion in the architectural field of CNC machines and lately of the industrial robots. And furthermore, this need finds its potential to thrive in digital machining. "*It is now possible to regard computer programming and architectural construction as conditional upon each other, and to see their reciprocity as fundamental to architecture after the digital age. As a consequence, the digital becomes concrete and tangible. Hereby the robot is both symbol of and tool for a profound reformation of the discipline*" [9]. The uptake of the innovation coming from new computational tools, together with the development of tailored fabrication machineries allows to directly engaging properties and features of matter since the early design phase. Innovative digital technologies redefine the relationship between design and construction, encoding new ways of thinking architecture. Considering the renewed position of architect after the digital turn, the designer is not only "*a passive observer of determined systems nor a determined manipulator of passive material, but rather, the manager of an unfolding process*" [10] and of an innovative digital chain. This practice goes beyond the boundary of complex computational thought, achieving the crucial architectural condition: its physical materialization.

2 Digital Fabrication in Architecture

"*Designing architecture isn't an activity that can be reduced to performance optimization or simulation of parametric evolution—it is a manifold cultural production*" [11]. Certainly digital computation extends the architect's human capabilities; nevertheless it fails to totally enclose the architectural problem. The advent of "*digital materiality*" [12] shakes up the research and triggers a fundamental rethinking of construction in architectural design. "*The Post-Parametric culture illustrates a shift away from the uncritical acceptance of parametric technology*" [13] and trend. The renewed focus on building materials and on the development of a digital fabrication craft restores the keynote that architecture basically is a material-based practice. It denotes that "*the close engagement of materiality is intrinsic to the* [digital] *design process*" [14]. Recently many CAM tools have spread, such as CNC milling machine, laser cutter, 3D printer or even industrial robot arm paired with special end-effectors. The increasing availability of these technologies does not only represent an astonishing technical advancement but it opens up groundbreaking approaches to design generation and implementation. Algorithmic design process can arrange the digital fabrication methods together with material properties. Therefore the digital model becomes the direct link between what is virtually represented (generated and optimized) and what can be physically built: architects are able to design digital workflows that merge concept and fabrication. The possibility to directly produce what is designed in bits-environment reveals we entered in the File-to-Factory age. In the last decade, design-to-production represents the modus operandi of architectural research, both academic and of some resourceful professional practices. Similar to computer in the 1980s, the cost of today's fabrication equipment is still prohibitive. However for the researcher the direct interaction with these machines is capital to learn how to use them and, above all, how they can enrich the design workflow. As Bonwetsch et al. [15] expressed, it is essential to apply the experience of digital production methods at the starting point of the design process, in order to relate the new possibilities to the previous knowledge. On this occasion we can draw a parallel between the early digital age and the digitally driven fabrication age: data paired with physical materials embody the matter for the File-to-Factory architecture researches, as well as information was the matter for the former computational studies. Coding becomes a design activity extended to the digital fabrication. Applying and programming tools offer a cutting-edge improvement to the exploration in architecture, going beyond the generation of abstract shapes and patterns. The new generation of CNC-machines are used to materialize the geometrical rules. A prime example is the spreading in architectural field of robotic arm: its high degree of kinematic freedom opens up real possibilities to directly fabricate, from the computational virtual models, highly performative projects.

Information and material are interwoven in "*digital materiality*" [12]. This digital pair suggests the idea that any building material is "*not just a passive receptor of form*" [16], but is itself a source of information for the design. Consequently the matter of architecture after the digital turn isn't the only information, but the bit-atom duo. The uptake of 6-axis—or more—industrial robots as the digital operating arms of architects, allows acquiring several new techniques of digital fabrication, mainly adopted from the industrial automotive field. The new processes of digital manufacturing—tessellating, sectioning, folding, stacking, cutting, contouring, milling, layer-based material deposition, and so on—rely upon the seamless digital integration from conception through to construction. This possibility of direct fabrication from the digital model enhances the depth of digital architecture. "*Materiality is enriched by the rules of the immaterial world of digital logics, such as its ordered nature or accuracy*" [9]. "*Digital materiality*" [12] turns out as the second fundamental rethinking in digital design age. The presented case studies meaningfully demonstrate that File-to-Factory age prototypes are freed from conventional representational function. The cornerstone of the presented experiences is their making approach. It finds its conceptual ground in the need of architects to control the very end of the overall design process: the fabrication.

3 Case Studies. Leading Robotic Fabrication Experiences: Between Conventional and Non-standard Building Materials

Digital fabrication enriches the simulated and optimized performance of forms with the sensorial, physical and material nature which is own of architecture. By means of new generation CNC-machineries, the model develops a new identity. One-to-one digital fabrication offers the way to digitally design architecture, far from the merely representation of form. It introduces an intermediate focus between experimental research and the actual building size, revealing the archaeology character of digital representation of forms, "*overwhelmed by new capabilities to fabricate, like never before*" [18]. Linking File-to-Factory attitude with powerful capability of algorithmic design—this latter provides tools which make able architects to weave together rules abstracted from several disciplines i.e. geometry, biology or materials science—leads the approach of several academic institutions, like the Institute of Computational Design (ICD) together with the Institute of Buildings Structure and Structural Design (ITKE), both based in Stuttgart. Since 2008 architects, structural, geodetic and biomimetic engineers, material scientists and biologists co-operate to test innovative architectural possibilities merging the computational and the physical in a digital fabrication synthesis. The ICD/ITKE Research Pavilion 2011 is based on a biomimetic approach for both the development of construction systems and generative computational design processes. It takes advantage from previous theoretical research for extracting morphological principles embedded within the sand dollar's skeleton—a sub-species of Echinoidea—related to structural and architectural demands as drivers in the context of performative morphologies in architecture. The skeletal shell of the sand dollar is a modular system of polygonal plates, which are linked together lengthwise the edges by finger-like calcite protrusions. High load bearing capacity is achieved by the particular geometric arrangement of the plates and by the density of the joining system. The design process leads to the plywood fabrication by means of novel computer-based design and simulation methods, along with custom-made computer-controlled manufacturing methods for its manufacture. Hence the tangible innovation consists in the possibility of effectively extending the recognized bionic principles and related structural performance to a different scaled up material-system throughout a continuous computational process. Acting in a seamless system, algorithmic code and digital fabrication strategy evolve together. A new robotic fabrication process (Figs. 2, 3) customized on the 7-axes ICD/ITKE's industrial robot—six revolute axes linked to an additional turntable—updates the traditional finger-joints used in wood carpentry (Fig. 4). The traditional woodcraft is characterized by manual and laborious operations. Furthermore, the most efficient wood connections usually require steel elements generating further problems, such as different temperature behaviour and corrosion. Here the efficient robotic fabrication of differentiated connections of the shell-elements improves the highly performative construction system based only on plywood

Fig. 2 © ICD/ITKE Research Pavilion 2011 robotic fabrication (*top left*)

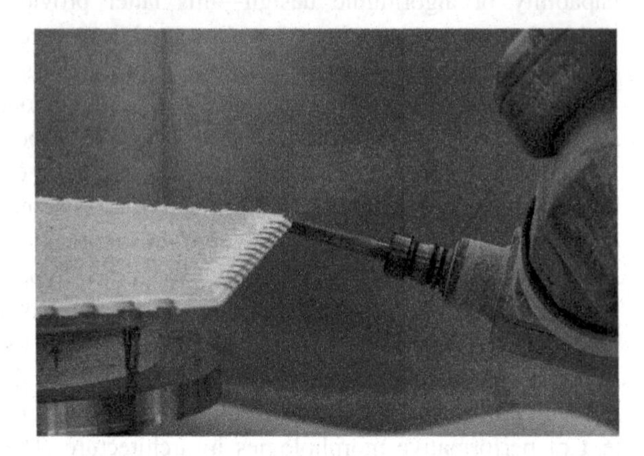

Fig. 3 © ICD/ITKE Research Pavilion 2011 cutting finger joints (*top right*)

plates, withstanding normal and shear forces without the need for additional fasteners. The new adaptive finger-joints system represents the core of the physical translation of the virtual model generated and optimized starting from the biological analysis. The final configuration of the interlocked plywood sheets demonstrates a high degree of adaptability and both spatial and structural performances due to the

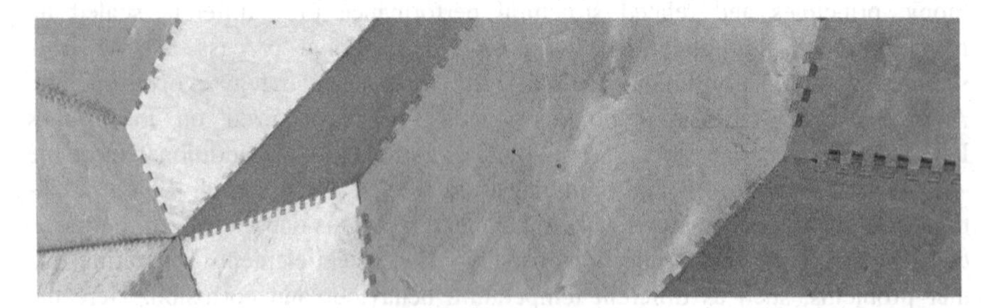

Fig. 4 © ICD/ITKE Research Pavilion 2011 closeup (*bottom*). All the images are used with permission

geometric differentiation and to the high precision ensured by the fabrication. Here a very large number of parts are combined, so the potential of digital design and production processes can be better exploited. Through the combination of the computational model and the robotic material production, the researchers realize a double layer shell of 6.5 mm (1/4 in.) thin sheets of plywood. Architects supported by structural and geodetic engineers script the form-finding rules to achieve a seamless digital chain from the digital model to the automated fabrication of plates. According to Schwinn et al. [12], designers determined the following robotic forming fabrication sequence taking in account geometric limits and particular machine constraints (i.e. securing a collision-free processing between wooden plate and spindle of the end-effector). First, milling the polygonal plate's outline; second, milling the edge's miters; and finally indenting the finger joints, using the tip of the milling bit. A necessary robot path simulation is the last step before the CNC code extraction: the fabrication's simulation allows to check the transformation of Cartesian coordinates to the angular coordinates of the machine's revolute joint space, needed to shape about 100,000 individual finger joints [19]. The direct communication between digital model and the 7-axis robot occurs through the G-code: exactly one fabrication file is generated per plate so that the robot is informed to physically contours the more than 850 geometrically different components of plywood (Fig. 5). The formed plywood sheets are assembled by hand obtaining the several double-layered cells; then they are placed and bolted together. Because of the lightweight, the pavilion is also secured to the ground to prevent it

Fig. 5 © ICD/ITKE Research Pavilion 2011 view north east. Used with permission. The highly performative construction system of this lightweight pavilion shows the robotic's potentiality in mono-material structures

from being lifted up by the wind. The explorative process of this Research Pavilion represents an integrative approach to architectural design with a special focus on how the robotic fabrication strategy can act as a design driver for plywood morphologies.

Fabrication capabilities expanded by the technological developments stimulate architects to rethink the way to use conventional building materials and building elements as well. Since 2006 Gramazio & Kohler Research, professors at ETH in Zurich, kicked off their investigation pointing out the dynamic role of material in dialogue with the robotic fabrication. One of their on-going researches faces new adding-on logics of bricks, one of the oldest, quasi-standardized building element known to man and most widely used both as structural or as coating element. On one hand first experimentations—using clay-bricks and wooden bricks—might appear unripe such as Structural Oscillation (2008, Venice), The Programmed Columns 2 (2010, Zurich), Stratifications (2011, London), Echord (2011–2012, Zurich) or the leading Facade Wineyard Gantenbein (2006, Fläsch). On the other hand these projects introduce robots in the architectural fabrication process in a massive way (Fig. 6), using them for their main features: the precision, the customized repetitiveness and the wide availability of degrees of freedom (both in movement paths and in end effector customization). While the human thought sets various parameters, makes design and structural decisions, the robot is set to shape

Fig. 6 © Gramazio Kohler Research, ETH Zurich pre-fabrication steps of the modular wall of the leading Facade Wineyard Gantenbein at ETH Zurich. (*Left*) Used with permission

the material in many different ways. Furthermore the team implement the in situ logic, exceeding the limitations of standard brick construction: the teamwork is able to program and produce highly specific building elements, which otherwise they could not be built manually. This co-operation starts to give authority to the robotic fabrication in architecture initiating a novel aesthetic and structural understanding. In 2007–2008 Gramazio & Kohler have got the idea to house an industrial robot within a modified container combining the advantages of prefabrication with the advantages of in situ fabrication: it's R-O-B.

ROB's Mobile Fabrication Unit presents an industrial robotic 6-axis arm that can be equipped with different end effectors able to grab, glue, extrude, weld, bend or mill materials. Gramazio & Kohler's design logic does not establish a dichotomy between the human design activity and the robotic fabrication, but a collaborative process. Architects design geometrical and structural rules and they formalize these rules within an algorithmic environment to control the entire procedure and to have the direct possibility to communicate with machineries. Then the robot places each brick in its individual position, but it simultaneously performs other operation, like applying glue: these two steps are crucial because all the overlapping is different and both contribute to the "*formation of the three-dimensional tessellated wall units*" [17]. Not only. The design experience Stratifications is an installation that evolves from the automated overlapping logic. Here the robotic arm is used to assemble multi-thickness wooden bricks and it is equipped with a particular end effector: it does not only grab bricks. It has scanning sensors, which receive stability information from the positioned bricks, so the machine can re-establish the stable position of each brick, if labile. This kind of relation between design, machine movements, fabrication and sensor-driven operations are in loop and they affect the subsequent operation. The model and, above all, the material inform the construction in real time. "*The constant prototyping of robotic brick-laying strategies makes them a great push towards real industrial robotic construction. However, they still remain constrained to the laboratory or to very controlled outdoor environments. This, together with the fact that robots are not adept at working in unpredictable environments*" [20], offers the occasion for further development for on-site fabrication. A more recent challenge is the Mobile Robotic Unit, which is an industrial robot fixed on a mobile unit and equipped with 3D sensors that inform the robot on its position allowing it to self-calibrate according to the surrounding [21]. This approach adds more flexibility to the previous robotic experiences of on-site fabrication, developing a more compatible system with the dynamic nature of a construction site.

In the course of the current digital paradigm shift, the robotic fabrication points to the use of non-standard and non-discrete building materials. It embeds conventional manufacturing technique and, by that, it expands the performance capability of digital manufacturing in architecture. A variety of structural morphologies can evolve thereof, not only providing the well-known optimization of building efficiency through the digitally controlled manipulation of material, but also allowing for a convergence of material aesthetic and structural concerns. The on-going challenge Mesh-Mould (2012–2016, Zurich) achieves innovative concrete building solution employing a robotic arm's toolset, updating the ferrocement system originated in the 1840s in France. This research is a Provisional Patent Application and best expresses the plastic feature of concrete, expanding the customization of robotic fabrication. It focuses on the improving of some weak points of the reinforced concrete. The reinforced concrete is one of the most common building systems used in architecture. It has the trait of a labour-intense process, unsuitable because it produces a lot of waste and it has high cost of form. Mesh-Mould (Fig. 7) tries to automate the building process, it lowers costs of form and is a waste free process. The first detailed 3-dimensional extrusion of ABS (Acrylonitrile Butadiene Styrene) shows how to combine formwork and reinforcement into one single robotically fabricated construction process, overtaking the traditional process. The degrees of freedom of the Mobile Robotic Unit allow producing the exact designed shape conforming the spatial reticular structure according to the optimized digital model (Fig. 8). So, the 3D printing fulfils the differentiated density in accord with the properties of the reinforcement configuration and forms the scaffold for the concrete casting. After extrusion, concrete is poured into the reticular mesh, covering the whole. In situ fabrication allows the overcoming of traditional paradigm of prefabrication, while it further develops the digital fabrication process itself. When implemented with strong materials, this on-site approach clearly eliminates the issue of structural weakness in construction

Fig. 7 The Programmed Columns 2 (2010) at ETH Zurich

Fig. 8 © Gramazio Kohler Research, ETH Zurich prototype of mesh-mould

joints or size restrictions concerning the transportation of prefabricated components [22]. In fact, this early stage of the Mesh-Mould prototype development is enriched by an on-going research direction that looks at the opportunity of implementing this fabrication logic with the use of rebar and also at a larger dimension. The potentiality of digital fabrication process offers also the opportunity to experience and test non-standard building materials. The ICD/ITKE Research Pavilion 2012 and Research Pavilion 2013–14 are two cutting-edge projects, which merge the customized prefabrication advantages and the modular logic together with innovative and unusual building materials. Both highlight the use of fibre-reinforced polymers (FRP) in order to achieve the robotic fabrication of a lightweight structures. The performative nature of FRP system is further enhanced through design considerations transferred from the biological field that allow for a higher morphological articulation in lightweight constructions. The Research Pavilion 2012, with its reduced objective in regard to program and permanence, investigates the robotic filament winding in order to explore construction-oriented innovations and to efficiently transfer the exoskeleton's morphological principles of arthropods. The

Fig. 9 © ICD/ITKE University of Stuttgart Research Pavilion 2012. Fabrication process

pavilion results as a lightweight and self-supporting structure fabricated as a monocoque construction. The Pavilion is made of translucent glass and black carbon fibres (Fig. 9); the flexible strands are coated in epoxy resin before being wound around a temporary metal scaffold, where they hardened and they give support each other. The fibres are spanning between defined anchor points on the frame, which control fibre position and allow for a continuously changing fibre orientation. About 130 h of robotic production time are totally automated; a custom tool on the robot's 3.9-metres-span arm precisely positioned the 90 km of fibres on the scaffold (Fig. 9), which is fixed on a turntable, avoiding the typically required mould or custom formwork [23].

The FRP fabrication technique is further tested and expanded within the Research Pavilion 2013–14: the main aim is the development of a coreless winding technique for modular, double layered fibre composite structures, which reduces the required formwork to a couple of polygonal frames while maintaining a large

Fig. 10 © ICD/ITKE University of Stuttgart Research Pavilion 2012. Overall view. All the images are used with permission

degree of geometric freedom (Fig. 10). The possibility of retrieving in detail the biological principles for the adjustment of the innovative architectural response has offered the engineers and architects involved in the project powerful means to parametrically design the very same component by abstracting fundamental geometrical decisions since the early design stage e.g. the thickness and the running direction of the fibres. In these last two cases the computational design, the robotic fabrication simulation and finally production specifically allow the development of a highly performance structures. The FRP differentiation coupled with the twist of the two anisotropic fibres lends a solution to the structural problem [24]. These two Pavilions suggest a new aesthetics and they materialize the abstracted morphological model into extremely new lightweight efficient structures. Summarizing, these disclosed experiences show how computation interlaced with automated custom robotic fabrication update the digital approach that inherit from the Nineties the logic of architecture as a process that looks for the "*goodness of fit*" [2] between ideas, design rules and materiality. With the digital uptake of fabrication techniques —milling, mould-making, winding and casting, among many others derived from traditional crafts—the architectural issue is facing the mechanisms of digital materialization: the computational field, which orders and simulates patterns, becomes a physical material order (Fig. 11).

Fig. 11 © ICD/ITKE University of Stuttgart Research Pavilion 2013–2014: making of the doubly curved glass fiber geometry and carbon fiber reinforcement. Used with permission

4 Conclusions

The current computational attitude lays its roots in the late Eighties and digital can be defined a cultural achievement resulting from centuries of human engagement with logic. The endless spread of computer, as a multifaceted and fascinating tool, motivates architect to exploit the human potential for associative thinking in order to discover new organizing principles.[5] The digital turn in architecture promotes computer as the new design logic; the current interest is on the development of material-oriented designs and in general establishing new relations with the physical dimension of architecture. "*The manifest form [...] is the result of a computational interaction between internal rules and external morphogenetic pressures*" [25]. As Kilian states, "*it is of less interest to use technology to merely solve problems of complexity in construction that were previously impossible to build*"

[5]The main feature of computer language is that it is an iteration of procedures: as an a priori schema, based on Alexander's logic diagrams, digital design shapes a potentially endless evolution. Digital culture lends a scientific aura to design practice, so architectural design is both deterministic and continuous. However, when researchers acquire digital logics as one's own, they build up a system, which is totally different from the reductive use of software to draw architecture.

[15] because of lack of appropriate tools. Nowadays, the melt of computation and digital fabrication identifies a new scenario: an open-ended design process, which combines the virtual form-finding with the material feasibility. The watchword "*Material is informed*" [12] isn't only a slogan, but it becomes an operative way to exploit the potential of computational design and, above all, the digital production process. The proposed case studies underline the digital connections between form, function, materials and digital fabrication facilities. For architecture, the relevant aspect of CNC-machines and particularly of industrial robot arms is the serial production of bespoke elements; their application definitely broadens the automation in architecture. However the computer-numerically controlled phase of the construction process is only a part of the whole effort necessary to fabricate any architecture. The information acquired from materials inform the design: this fundamental criterion of craftsmanship process is now digitally translated and updated, at the same time. We are witness of the spread of architectural designs that disclose the information stored inside the material across an aware implementation of the robotic fabrication logic throughout the entire architectural process. Digitally built architecture, provisional or not, encompasses a much larger field of interest than the representational one: digital fabrication is a practical as well as a cultural revolution. In the File-to-Factory age the researchers rationalize geometrical strategies for synthesis between procedural data design and the act of making. In this context the form appeals to the senses, while continuing to declare their distinctly procedural derivation. Architectural forms reach a wider and manifold identity when translated in real matter and usable production. The materiality of computational works expresses the physical behaviour of the form: it implements the embedded capacities of matter into physical forms. Finally, the digitally fabricated forms act performance (i.e. structural, spatial, tangible, aesthetic) and they not only simulate them. "*Digital materiality*" [12] is the post-digital turning point and, at the same time, it reopens some issues. The implementation of physical properties shifts the role of the model further away from mere representation towards a manufacturing process. In order to go beyond the computational thought, we need to regard an all-inclusive approach to fabrication methods that effectively combine human thought, robotic manipulation, and material. The presented digital fabrication logics, with a new material awareness, allow the conception of an automated workflow, which arrange design and production within a seamless path. Research experimentations enhance the serial production settings of robot arms that are now used for tailored and manifold pattern of fabrication. In this way robotic arms, equipped with customized end-effectors, are able to build pioneering geometries and to explore new materials and techniques. This new digital craft opened by the synchronous spread of CNC fabrication tools and of a post-digital attitude, has yet to be improved to really influence the way we will build architecture at each scale.

References

1. Davidson, P., & Bates, D. (1997). Architecture after geometry. In Toy, M. (Ed.), *Architectural design* (Vol. 67, No. 5/6). London: Wiley.
2. Alexander, C. (1964). *Notes on the synthesis of form*. Cambridge, Massachusetts: Harvard University Press.
3. Thompson, D. W. (1959). *On growth and form*. Cambridge: Cambridge University Press.
4. Kwinter, S. (2003). *The judo of cold combustion*. New York: Raiser + Unemoto, Atlas of Novel Tectonics, Princeton Architectural Press.
5. Meredith, M. (2013). After after geometry. In B. Peters & X. De Kestelier (Eds.), *Computation works: The building of algorithmic thought, architectural design* (Vol. 83, No. 2). London: Wiley.
6. Kwinter, S. (2003). *The computational fallacy*. Thresholds no. 26. Wilmington: Kirkwood Printing.
7. Allen, S. (2011). Interview 1. In G. P. Borden & M. Meredith (Eds.), *Matter: Material processes in architectural production*. London: Routledge.
8. Negroponte, N. (1995). *Being digital*. New York: Vintage Book.
9. Gramazio, F. (2014). *Digital materiality in architecture*, speech during lift conference, February 2014, Genève.
10. Reiser, J., & Unemoto, N. (2006). *Atlas of novel tectonics*. New York: Princeton Architectural Press.
11. Menges, A. (2013). Performative morphology in architecture. *Serbian Architectural Journal* 5, Belgrade.
12. Gramazio, F., & Kohler, M. (2008). *Digital materiality in architecture*. Zürich: Lars Müller Publishing.
13. Pérez, S. (2011). Towards an ecology of making. In G. P. Borden & M. Meredith (Eds.), *Matter material processes in architectural production*. London: Routledge.
14. Booth, P. (2009). Digital materiality: Emergent computational fabrication. In *43rd annual conference of the architectural science association, ANZAScA 2009*. University of Tasmania.
15. Bonwetsch, T., Bärtschi, R., Kobel, D., Gramazio, F., Kohler, M. (2007). Digitally fabricating tilted holes. In *Predicting the future: 25th eCAADe conference proceedings* (pp. 793–799). Frankfurt am Main.
16. Pottmann, H. A., Hofer, A., & Kilian, M. (2007). *Architectural geometry*. Exton: Bentley Institute Press.
17. Iwamoto, L. (2009). *Digital fabrications architectural and material techniques*. New York: Princeton Architectural Press.
18. Lynn, G. (2014). *Archaeology of the digital*. Montreal: Canadian Centre for Architecture.
19. Schwinn, T., Krieg, O.D., Menges, A., Mihaylov, B., & Reichert, S. (2012). Machinic morphospaces: Biomimetic design strategies for the computational exploration of robot constraint spaces for wood fabrication. In *Synthetic digital ecologies: Proceedings of the 32nd annual conference of the association for computer aided design in architecture* (pp. 157–168). San Francisco: ACADIA, College of the Arts.
20. Keating, S., & Oxman, N. (2013). Compound fabrication: A multi-functional robotic platform for digital design and fabrication. *Robotics and Computer-Integrated Manufacturing, 29*(6), 439–448.
21. Helm, V. (2014). In-situ fabrication: Mobile robotic units on construction sites. *Made by Robots: Challenging Architecture at a Larger Scale, Architectural Design, 84*(3), 100–107.
22. Hack, N., Lauer, W., Gramazio, F., & Kohler, M. (2015). Mesh mould: Robotically fabricated metal meshes as concrete formwork and reinforcement. In *Ferro-11: Proceedings of the 11th international symposium on ferrocement and 3rd ICTRC international conference on textile reinforced concrete* (pp. 347–359).

23. Reichert, S., Schwinn, T., La Magna, R., Waimer, F., Knippers, J., & Menges, A. (2014). Fibrous structures: An integrative approach to design computation, simulation and fabrication for lightweight, glass and carbon fibre composite structures in architecture based on biomimetic design principles. *Computer-Aided Design, 52,* 27–39.
24. Dörstelmann, M., Knippers, J., Menges, A., Parascho, S., & Schwinn, T. (2015). ICD/ITKE research pavilion 2013–14—modular coreless filament winding based on beetle elytra. *Architectural Design, 85*(5), 54–59.
25. Kwinter, S. (2008). *Far from equilibrium essays on technology and design culture.* Barcelona: ACTAR.

Encoding Gesture-Oriented Human Behaviour for the Development and Control of an Adaptive Building Skin

Odysseas Kontovourkis, Kristis Alexandrou and Stavros Frangogiannopoulos

Abstract This paper discusses soft structural transformation strategies and responsive motional activation control methods for the development of an interactive building skin, responsible for regulating inner light conditions, with particular feedback from human position and stature. Inspired from the latest employment of smart mobile devices in building industry as an on/off wireless control switchboard, the present project demonstrates an alternative, gesture-based activation methodology towards a more natural and approachable way of controlling architectural environments. Elastically deformable planar elements of low thickness, in the basis of bending-active principles, have been chosen for the composition of a 'soft', modular adaptive building skin that enables flexible kinetic transformation with enhanced diverse geometrical conversion. Alongside the question of responsive motional control strategies, the search for a practical design setup is investigated aiming to establish a computational-oriented methodology that integrates information from both direct physical sensory data and indirect computational programming and organization of the input information. This is discussed through the case studies, where the physical performance of individuals is examined to generate input parameters responsible for the appropriate gesture activation of the digitally simulated responsive skin. User's physical actions are recorded via the synergy of multiple sensors and classified into two main categories. Computation of both data resulted on an enriched user motion decoding, and therefore the encoding of specific structural reactions.

Keywords Computational design · Adaptive architecture · Wireless gestural control · Bending-active members · Physical computing

O. Kontovourkis (✉) · K. Alexandrou
Department of Architecture, University of Cyprus, Nicosia, Cyprus
e-mail: kontovourkis.odysseas@ucy.ac.cy

K. Alexandrou
e-mail: kalexa01@ucy.ac.cy

S. Frangogiannopoulos
Michael Cosmas Architecture LLC, Nicosia, Cyprus
e-mail: stavros_esp@hotmail.com

M. Rossi and G. Buratti (eds.), *Computational Morphologies*,
https://doi.org/10.1007/978-3-319-60919-5_10

1 Introduction

Examples of adaptive architectural structures with dynamic spatial relationships through human involvement are numerous and variable according to the kind and type of adaptiveness. Some examples are referring to cultural adaptation, societal, environmental or aesthetical, concerning means of communication, architectural fashion, social interaction, solar exposure etc. [1]. Key element for the appropriate transformation of adaptable systems and the subsequent geometrical reconfiguration of their structural elements lies upon the right selection of the structural system and its potentials to reach a satisfactorily adequate conversion. Recently a number of research activities focused on the application of elastic materials in architecture [2]. Materials characterized by passive elastic properties can form mechanisms of autonomous, elastically reversible deformational behaviour. Compared to technical stiff catenary systems, bending-active mechanisms replace multiple local hinges with a single planar component, where the distribution of actuating force spread over a wider material area and deformation takes place through bending. This soft approach renders the possibility of a more complex—single or double curvature surface formations from straight planar elements through single, simple and low actuating force.

In this frame, elastically deformable adaptive structures that serve interactive dialogue with various external agents, give emphasis on the beneficial privileges that such systems may deliver to the design and the resolution of its kinematic performance. The REEF installation at Storefront for Art and Architecture, New York [3], demonstrates how the shape transformation of a system of 'petals' can result into a more sensitive architectural space experience. It employs 'soft' materials such as bending rods and membranes to yield its transformed shape with great fluidity. The kinesthetic of the structures reconfiguration is stimulated and predisposed by the pedestrian flow from the exterior surroundings of the building. Its overall performance highlights how the negotiation of public and private can produce a new perception of communication and expression. Similar approach that deals with soft material deformability has been undertaken by SOMA Architects in close collaboration with Knippers Helbig advanced engineering, for the synthesis of the kinematic façade of Expo 2012 in South Korea [4]. In this situation, the structural elements have been placed vertically and serve a dual responsiveness purpose; the regulation of the sunlight that covers the inner space of the museum and the expression of aesthetically pleasing kinematics through animated patterns of structural distortion. The distribution of actuating force is placed asymmetrically, forcing the vertically oriented lamellas to form a three dimensional (3d) deflection. The 'Softhouse' residential project in Germany [5], establishes an entirely ecological responsive approach, concentrating completely on stimulus derived from the ambient environment. It can also be classified as a performance-based design approach, as the element transformation increases the efficiency of the building on harvesting sunlight energy, while simultaneously provides sunlight shading during the summer period.

On the other side of this twofold approach comes the search for the right means of interaction [6], focusing on the source and type of the input information that enables structural activation performance. Spatial interactivity triggered by the user movement, can be synchronized out of a vast agenda of communication signals and tools responsible for translating physical information into computational-able data. Emergent technologies originally developed in parallel fields of architecture have seized the attention of researchers to deep into the potentials of structural activation and the magnitude of control that the user may achieve, as to dynamically interact with space, under a simple and highly responsive syntax of communication [7]. Technologies that allow users to interact with and control over adaptive structural arrangements can be classified into three main categories: touch or multi-touch gesturing, physical gesturing and cognitive control[1] [8]. Such technologies and processes are nowadays embedded into several smart devices and sensing artefacts,[2] cumulating the possibilities of interactive applications to come to fore. Smart mobile devices consist some of the most powerful sources of information exchange between the physical and digital environment. Due to their multiple sensors embedded and their sufficient computing power, smartphones make feasible the creation of a more precise and accurate translation of the physical movements into numeric digital data. Compare to press or touch interfaces of control used in the industry for home appliances [9], gesture-based interaction, which supports sensors such as gyroscopes, accelerometers and gravity sensors, can perform a more adequate depiction of the physical actions. This can enrich the emerged associations of movement with detailed parametric control including speed, distance, direction, orientation etc., minimizing the possibility of errors to arise during the computational design. In addition to this, identification of the user, which also comprises a critical aspect for future developments on this prospect, is guaranteed due to the fact that mobile devices are personal and encrypted with passcodes of finger blueprints.

Based on an initial work produced within the framework of a postgraduate course in the Department of Architecture at the University of Cyprus and presented in the 3rd eCAADe Regional International Workshop 'Computational Morphologies' in Milan [10], this extended paper demonstrates a further work dealing with the development and control of an adaptive building skin through the encoding of gesture-oriented human behaviour. This paper is divided into seven

[1]Control of cognition: Cognition has to do with how mobile devices (operating system) understand and act in respect to user's usage or preferences. It is defined as a set of processes and data examination that are purposely performed to activate a more interactive user-phone experience. Cognitive control is actually a mechanism of how digital platforms learn, recognise user's actions, remember choices or particular data and effectively suggest options or even make specific tasks automated. This step is developed over time according to user's activity and can manage both short- and long-term control [13].

[2]Sensing artefacts are sensing devices made by human being. These devices vary in technological manufacturing and nature of sensing. Many technological products are composed from sensing devices to provide users with valuable information regarding their status, operation or even failure. In most cases they work as protecting agents to prevent failure or to activate a special feature to avoid such incident to happen [14].

sections. The second part demonstrates analytically the general strategy for computational design development; the data classification, computing and motion acquisition process are described in section three and four respectively; the fifth section describes the overall adaptive building skin development and lighting performance results; and finally, section six and seven draws general conclusions and future prospects for the current research.

2 Strategy for Computational Design Development

The suggested strategy aims to establish a wireless sensing communication between the user and the digital platform of operation. This is done, firstly, by programming the input data derived from sensing devices and, secondly, by defining the geometrical rules that govern system's transformation behaviour (structural deployment). Due to the increased amount of material and servos needed to implement such an installation in the physical world, it has been decided to digitally model and simulate the soft building skin and to physically fabricate and assemble only a single module of the system. The whole procedure has been modelled in the parametric modelling software Grasshopper[3] (plug-in for Rhino).

Analytically, in a preliminary stage, the transmitted values from the input sensors are collected and examined, aiming at formulating a computational language of communication and response between the user's motion and the digitally defined geometry. In a further stage, the structural system behaviour is examined by introducing a 'soft' material with low bending stiffness properties, having as major characteristic a reversible geometrical deformation. This is done, firstly, to observe actual structural transformation capacity and apply appropriate geometrical rules to the digitally simulated model based on this preliminary physical prototype. This investigation is interrelated with geometrical constraints and interdependencies, in order to depict the physical structural distortion as truthful as possible. As it has been mentioned, a parallel investigation of the system is examined in real scale by developing a final single module of the physical prototype. This is fabricated and assembled to verify and validate the correlations of its physical behaviour with the digital model replica and to demonstrate the activation behaviour of the suggested building skin using actuation mechanisms that control system's tunings (Fig. 1). The overall digital building skin as well as the final physical prototype are combined and compared with the proposed strategy (Fig. 2) to achieve a dialectic

[3]Grasshopper is a parametric plug-in for Rhinoceros 3D. It allows custom programming for existing commands of Rhino as well as custom definition of commands using programming languages. Several sub-plugins have been formulated by developers which include pre-set components to enable communication between other official applications or hardware components such as sensing devices (Kinect camera etc.).

Information about Grasshopper at: http://www.grasshopper3d.com.

Grasshopper plug-ins at: http://www.food4rhino.com/?etx.

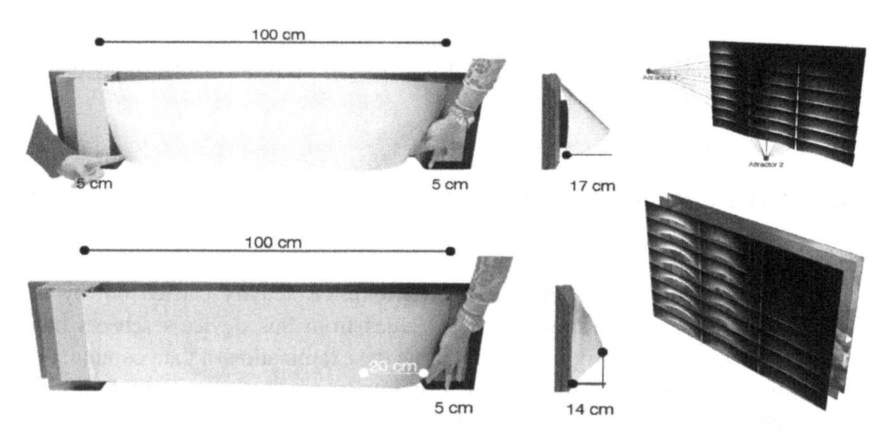

Fig. 1 *Left* Maximum bending deformation of the physical prototype. 8 cm inwards results in 17 cm of dispacement at mid-span. *Right* System: simulated bending behaviour of all modules of the system. Bending deformation varies according to users position in space

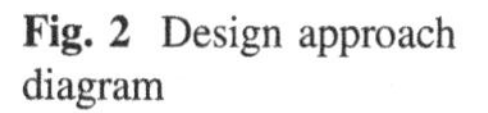

Fig. 2 Design approach diagram

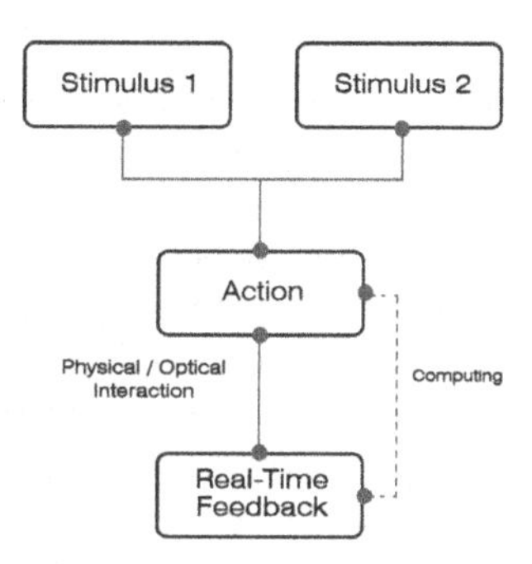

relation with the environment influencing the digital and physical development of the system in various ways [11].

3 Data Classification and Computing

3.1 Input and Output Devices

3.1.1 Microsoft Kinect Camera

Preliminary series of case studies are undertaken to enlighten the project with valuable sensor information regarding human position or movement in an interior space. A Kinect camera has been assigned as the position tracking sensor, tracing exclusively a single point of the human body and in particular, the point that represents the head of each individual user. The input information derived from the camera device are transferred to Grasshopper environment and are distinguished into point coordinate values that control the physical transformable behaviour of each module (louvre) of the building skin.

3.1.2 Android Mobile Device

An Android mobile device is employed to act as an orientation receiver (orientation sensor) and subsequently as the identifier of explicit activity carried out by the manipulator (linear acceleration sensor). Values from this device's sensors have been collected independently in order to allow the examination of data combination possibilities, the degrees of freedom or restraints (domain range) faced during the programmatic synthesis of the streamed information.

3.2 System Control

The system is controlled through an algorithm applied in Grasshopper. Forming a constant feedback loop of input and output information illustrated in Fig. 3, its role is to continuously collect data from all sensors manipulated by the user, then to analyze and process them in order to translate the values collected into digital transformation of the building skin and in parallel into physical motion behaviour of the servo motors, which control the final physical prototype alteration.

In order to achieve successful control of both, the digital skin system and the physical prototype, at a first stage, a secure and reliable connection between the sensors and the main console or computer are investigated. The Kinect SDK is used

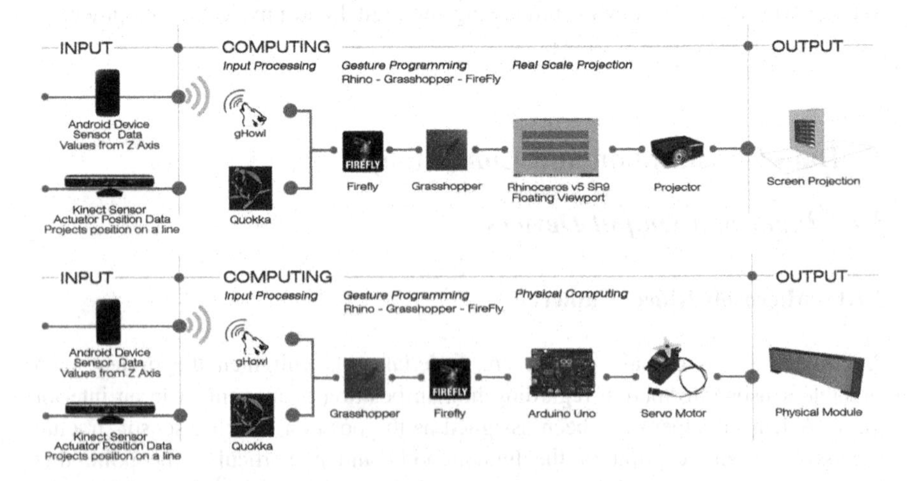

Fig. 3 *Top* Input processing for digital output. *Bottom* Input processing for physical output

as the driver responsible to recognize the Kinect camera's connection with Windows, and Quokka (plug-in for Grasshopper) is responsible to establish the constant communication and real-time data feedback between the sensor and the processing unit. The gHowl (plug-in for Grasshopper) is another key software, which undertakes the role of recognizing the IP Address of the Android mobile device and listening, via Wi-Fi, for values produced by its orientation and acceleration sensors. Finally, Firefly (plug-in for Grasshopper) is used to control the motion behaviour of servo motors that accelerate the transformation of the single physical louvre.

3.3 Physical to Digital Data Computation

The information derived from preliminary physical prototype deformation behaviour contain two main observations that are key features of the system alteration: first, the maximum tilt of the bending element being 34° from the vertical axis, and second, the two bottom corners of the bending louvre, which relocate inward at a maximum of 8 cm (Fig. 1). This data is correlated with the data values produced by human behaviour, aiming to control the digital transformation behaviour of the system. Specifically, the values captured by the mobile device determine the bending behaviour of each module, whereas the value captured by the Kinect camera device is responsible to control the distribution of the deformation of the individual modules and therefore, the area of influence of the overall system. The three dimensional tracking point of the user in space, defined by xyz values, determines the magnitude of deformation as well as the selection of the individual louvres that will be affected. For instance, the closer the user is to the system, the greater the deformation of the louvres is, whereas, the location of the user in space, influences proportionally the closest modules of the system. Data captured from the mobile device's multiple sensors are remapped and set to operate as additional factors to the already defined deformation distribution, primarily handled by the Kinect camera. Y values collected by the orientation sensor, that vary from 0° to 360° degrees, are set to adjust the magnitude of the bending behaviour of the system, and therefore, of the affected louvres. The physical control is achieved by the user while moving or maintaining his/her position in space. The value of 0° degrees appears while facing the façade and 180° degrees while having his/her back facing the system. When the value is 180°, the deformation magnitude of the system is decreased by 25% whereas the value of 0°/360° degrees leaves the system's deformation unchanged. At last, linear acceleration sensor data define the acceleration of the user in space. The specific value counts from 0 to 45 units, whereas 9.89 equals gravity acceleration (g). Out of a series of tests performed it has been observed that walking detection amounts from 2.15 to 2.75 units, while fast walking or sudden moves counts from 3.75 to 5.0 units. For the case of this research development, accelerometer is suggested to work as a true or false enabler instead of an analogue one. An increase in acceleration up to 3.75 units is assigned

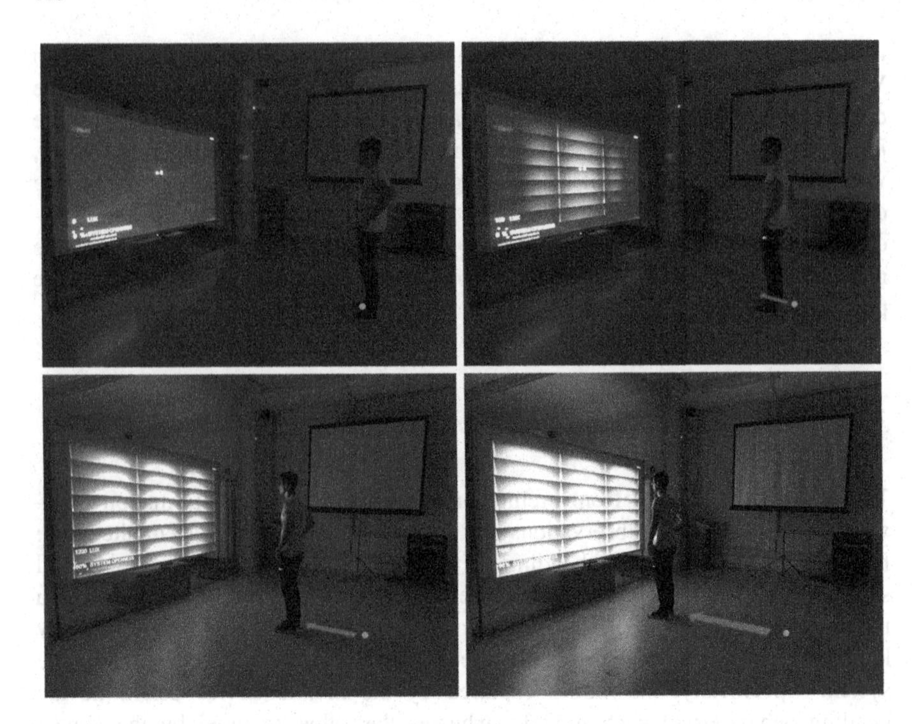

Fig. 4 Location tracking and system transformation reaction

to 'true' which enables the system to overcome all input data and apply 100% openness in all modules. This is decided considering the fact that sudden moves indicate a hurry and therefore a request for maximum visibility and brightness.

Due to the complexity of the parameters that rule the transformation of the building skin, a special classification of data is applied to the system to fit more appropriately favourable motions performed by the user. Attractor point, which is considered as the leading stimuli of influence, is separated into two modes of control. In this respect, 'standing and sitting' modes are introduced. Point constitutions (xyz) are decomposed and z values are isolated. Once z value reaches 1.35 m in height, a redefined practise of sensor data is applied, responsible to manage gestures that are more suitable to the sitting mode or the standing mode respectively (higher than 1.35 m) (Figs. 4, 5).

4 Motion Acquisition

As a starting point, motion activity of users in space can be distinguished into two main areas. Activities performed while sitting, such as reading, watching T.V. or even working, and activities performed while standing such as interacting with people, working or moving towards a different space. Observing such activities, it

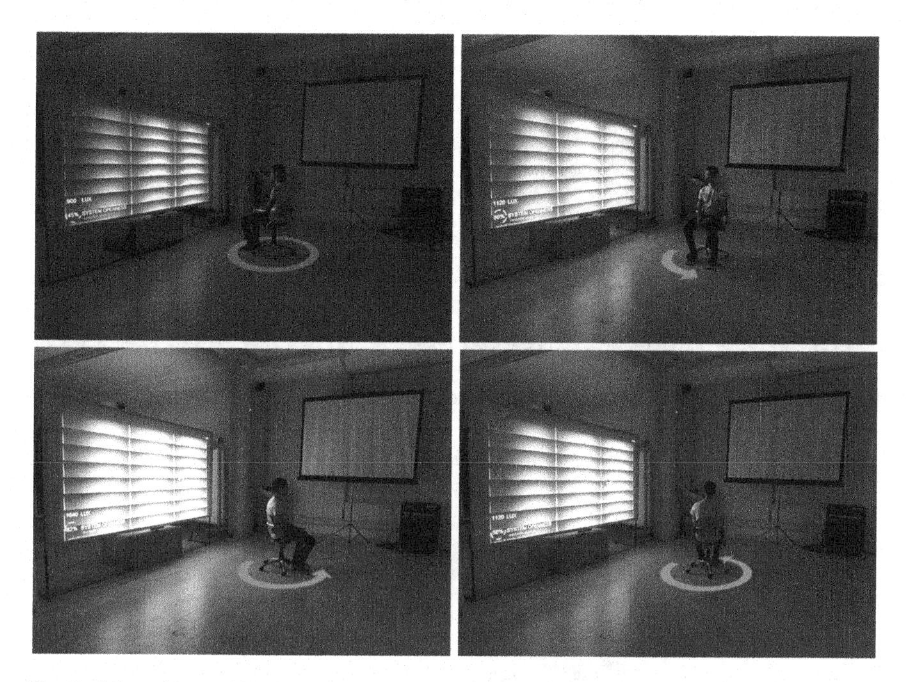

Fig. 5 Orientation tracking. User facing the system enable 75% of the actual system deformation whereas user facing the opposite direction enables 100% of the system's openness

can be realised that each one can be characterised or defined by different physical gesturing parameters. For example, acceleration, orientation or often position allocations in space are valuable indicators that can be utilised to decipher standing activities. Facing direction or revolving in space are physical body gestures that are part of the specific action that is currently active. In this respect, this research demonstrates several actions that can directly influence and control over the system's transformation.

Users in standing mode have a proportional effect on the surface of the simulated skin, according to their overall perpendicular distance. In addition to this, the area of influenced louvres is focused to the respective perpendicular position of system. Acceleration detection enables a 100% openness to the system as a signal of hurry or direct need of maximum brightness in space. During sitting mode, system response is updated with the parameter of orientation. User's body direction in space adds an additional filter to the overall deformation percentage of louvres as to adapt into current needs. For instance, a person looking the opposite direction of the systems location, an additional 25% of openness is directly added to the system, allowing more light to pass through, whereas once the user is facing the louvres, a 25% decrease of system openness is applied respectively. This is empirically assumed after several tests and results evaluation. In all cases, the mobile device is placed into the user's pocket, in order to allow physical gestures to be decoded and transferred to the digital interface. Indicative results of this process can be seen in Fig. 6.

Fig. 6 Random user activities and system's reactions accordingly

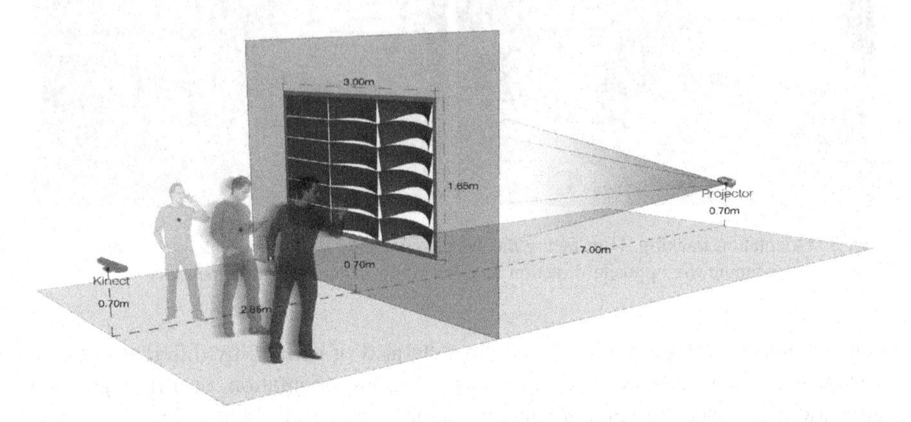

Fig. 7 Digitally projected system of louvres

5 Adaptive Building Skin Behaviour and Lighting Performance Results

The physical behaviour simulation of the building skin is achieved by establishing a projection screen in actual scale, where the transformation of system and its reaction with the users is displayed in real time. It is assumed that the investigation takes place in the interior space of a building, and in front of a facade with dimensions $3.00 \times 2.15 \times 3.00$ m that is facing south (Fig. 7). Purpose is to achieve an adaptive relation between the building skin and the users of space through wireless gesture control. By correlating skin's openness with lighting performance, the users are able to change the system's state and observe the percentage of illumination in real time. Also, they can make the necessary adjustment to improve inner lighting conditions.

Information in relation to the lighting performance of space is derived from Ecotect analysis software which records natural illumination (in lux values). In this

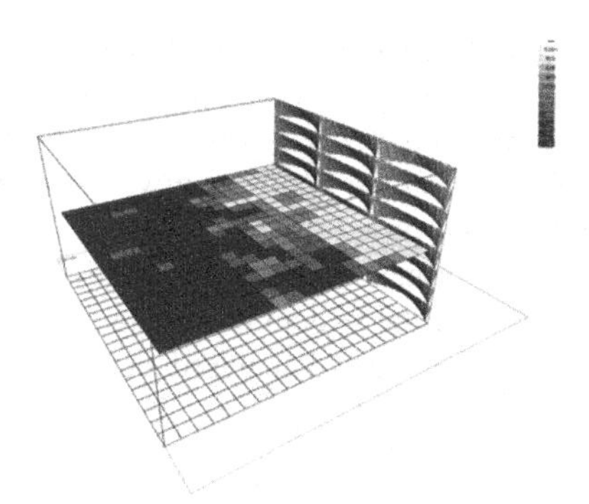

Fig. 8 Ecotect light simulation results (Openness—Illumination). Maximum def.: 100%—2508 lux

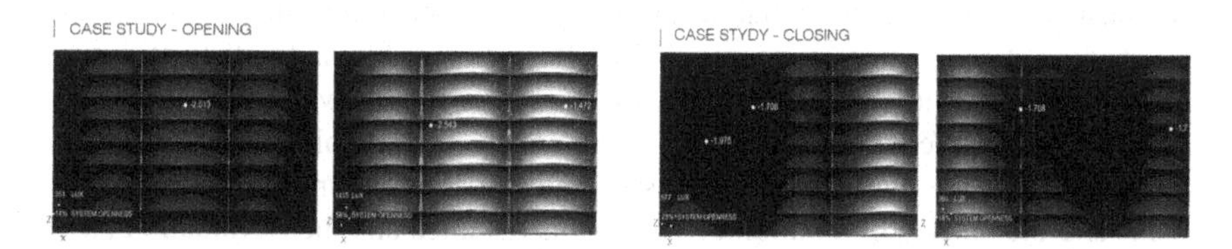

Fig. 9 Possible results of system's adaptive behaviour. *Left* System reaction enables opening of louvres. *Right* Enables closing of louvres

analysis, natural lighting levels are not date or time dependant, they represent the worst-case design conditions based on an 'average' cloudy or uniform sky distribution in mid-winter and the calculations are based on the split flux method [12]. Through an initial simulation, a proportional relation between openness and illumination is suggested, which relates the maximum openness (100%) with the maximum internal average annual illumination (2508 lux) (Fig. 8).

Following figure (Fig. 9) shows a number of patterns generated by the interaction between users and the building skin, together with information related to lighting performance. Initial results demonstrate the adaptive behaviour of the system, which is not part of an on/off situation, but of a smooth deformation procedure where proportional changes emerge, depending on users' activities. This, in combination with the selection of the particular elastic material, increases the responsiveness of the system, allowing a large number of results to be obtained.

Results show that the relationship between lighting performance and openness percentage is not directly proportional, but depends on the generated geometry. Furthermore, it is observed that the specific configuration and placement of bending modules (louvres), which are facing south, results in high insolation in the range of 2508 lux.

Motion acquisition suggested in this research resulted into a unique experience in space that is directly linked to a playful and adaptive light regulating effect. All physical gestures applied have been verified and resulted into vast and high resolution reactions of the simulated skin. In (Fig. 6) a record of random user activities is presented.

In a further stage, correlations between the results obtained through real time gesture control and analysis using the Ecotect software might be investigated. Aim is to verify the validity of the proposed strategy to be used for the design of building skin, optimizing lighting performance in interior spaces. Towards this direction, a number of geometrical states of the system according to their openness percentage (8, 14, …, 100%) can be made static and analysed, demonstrating respective natural illumination results.

6 Conclusions

In conclusion, the computational design strategy applied shows that the control of the shading device using personal mobile phone is feasible and sufficiently manageable. The way in which the projection screen is interactively related with the human behaviour concludes that the continuously changeable surface might be acting as a reactive platform of expression and visual communication between the inside and the outside of a building. The results obtained and their evaluation based on lighting performance criteria allows an investigation towards adaptive solutions, which might be characterized as optimum due to their feedback loop control mechanism introduced based on users' personal desires.

Finally, the wireless interaction with dynamic spaces reveals future potentials, in the way the regulation of architectural space and morphology is manipulated by the user's activities in real time. The integration of multiple sensor data from mobile devices and position tracking sensors within the architectural space, can formulate the experimental framework towards motion acquisition.

7 Future Prospects

Nowadays, the large amount of possibilities offered by the mobile devices is highly acknowledged and practically verified in numerous applications in multiple fields. The critical point of interest that allows these devices to succeed and to power on a broader area of research applications is based on the amount of information they can collect, exchange and process, in combination with the ease of portability they provide. In extent, multiprocessing and multitasking capabilities enable a fast and flawless rendering of the objectives carried out. Mobile phones are intentionally becoming the second half of our everyday activities and lifestyle, offering an easier, faster and more enjoyable way of experiencing everyday tasks and objectives. Recently, a number of revolutionary devices arrived in the market that can collect supplementary source of information such as body temperature, humidity and health-related information. Latest generations of smart devices, show a gradual embedding of these sensors into our body, otherwise announced as 'wearables'. In this respect, we may assume that these physical data collectors will be an inseparable part of our body, recording and translating many significant and precise information of our motional activities, feelings or even needs. In this perspective, a new way of user-space interaction may emerge, able to utilise this valuable data to provide more expressional, functional or comfy inner space conditions, bringing up a unique and inter-disciplinary way of dealing with the design and architectural synthesis, in regard to responsive and adaptive characteristics of space.

Acknowledgements The current paper presents an extended work that includes further research developments of a project initiated within the framework of the postgraduate course ARH-522—Advanced Computer-Aided Design Topics taught in the Department of Architecture at the University of Cyprus during the Fall semester of 2014. The initial investigation has been presented during the 3rd eCAADe Regional International Workshop 'Computational Morphologies' that was held in Polytechnic University of Milan on 14–15 May 2015 under the title 'Adaptive building skin development and wireless gesture-oriented control' and was authored by Odysseas Kontovourkis, Kristis Alexandrou, Constantinos Vassiliades, Stavros Frangogiannopoulos, Neofytos Dimitriou, and Christina Petrou [10]. We would like to sincerely thank all contributors participating in the previous part of investigation of the current extended work.

References

1. Schnädelbach, H. (2010). Adaptive architecture: A conceptual framework. In *Media City, Weimar.*
2. Lienhard, J., Lpermann, H., Gengnagel, C., & Knippers, J. (2013). Active bending, a review on structures where bending is used as a self-formation process. In *Space structures, multi science.*
3. Reef Installation. Retrieved from December 22, 2014 http://www.archdaily.com/29174/reef-an-installation-at-storefront-for-art-and-architecture
4. Knippers, J., Scheible, F., Oppe, M., & Jungjohann, H. (2012). Bio-inspired kinetic GFRP-façade for the Thematic Pavilion of the EXPO 2012 in Yeosu. In *I SS-PCS-symposium, Seoul.*
5. Softhouse Residence in Germany. Retrieved from December 22, 2014. http://arcintex.hb.se/uploads/images/pdf/Soft-House-Prospectus-2014.pdf
6. Dubberly, H., Pangaro, P., & Haque, U. (2009). On modelling what is interaction? Are there different types? 69–75.
7. Fox, M., & Polancic, A. (2012). Conventions of control: Catalogue of gestures for remotely interacting with dynamic architectural space. In *Association for computer aided design in architecture, San Francisco.*
8. Kemp, R. M. (2008). Interactive interfaces in architecture: The new spatial integration of information, gesture and cognitive control. In *ACADIA 08: Silicon and skin-biological processes and computation, Chiang Mai.*
9. Shahriyar, R., et al. (2008). Remote controlling of home appliances using mobile telephony. In *Smart Home, USA.*
10. Kontovourkis, O., Alexandrou, K., Vassiliades, C., Frangogiannopoulos, S., Dimitriou, N., & Petrou, C. (2015). *Adaptive building skin development and wireless gesture-oriented control [PowerPoint slides].* Presented at the 3rd regionsal international workshop 'computational morphologies' at Polytechnic University of Milan.

11. Kontovourkis, O. (2013). Physical data computing in adaptive design process. In *Proceeding of international conference of adaptation and movement in architecture, Toronto, Canada.*
12. Larke, J. (2011). *Energy simulation in building design* (pp. 266–268). London: Routledge.
13. Kieras, D. E., & Meyer, D. E. (1997). An overview of the EPIC architecture for cognition and performance with application to human–computer interaction. *Human-Computer Interaction, 12,* 391–438.
14. Artefacts: Oxford Dictionary: http://www.oed.com/view/Entry/11133?redirectedFrom=artefact

Interaction and Forming, How Industrial Design Is Changing

Maximiliano Romero

Abstract Present paper describes briefly historical excursus of the model making in product design and it development thanks to digital fabrication technologies. A critical analysis of design process is presented and related with application experiences in digital fabrication, co-design and interaction design. Some conclusions about the impact of this changes in the Italian design context are presented.

Keywords Interaction design · Physical computing · Computational design

1 Introduction

We are facing at what many call "the new industrial revolution" [1]. An historical change where the logic of production, distribution and consumption of material goods that we know may be redefined by the use of technology in a different way. At the basis of this change there are several factors, but certainly one of the most important concerns the possibility to easily exchange the necessary information for the products construction through the Internet. This information may relate to know-how and constructive experiences but also, and basically, the mathematical description of the component of these products.

Thanks to computer archives structured according to predefined encodings (files), is possible to exchange 2D and 3D models that can control automated machine tools, using the technology called Computerized Numerical Control (CNC). We are talking about building technologies industrially available for at least 20 years but which today arouse great public interest because available in small sizes, called Desktop, that are suitable for final user and not only for companies; we are talking about Desktop Digital Fabrication. "*Widespread access to these technologies will challenge the traditional model of business, foreign aid and education.*" [9, p. 43].

M. Romero (✉)
Department of Design and Planning in Complex Environments,
University IUAV of Venice, Venice, Italy
e-mail: mromero@iuav.it

M. Rossi and G. Buratti (eds.), *Computational Morphologies*,
https://doi.org/10.1007/978-3-319-60919-5_11

Fig. 1 Wooden model with dime of the Grillo phone, Marco Zanuso and Richard Sapper for Siemens, 1965 photography: Federico Pollini—Archive Giovanni Sacchi

What once was made by hand with wood or other easily shapeable materials, like the famous maquettes of Giovanni Sacchi (Fig. 1) for great designers of the 60's, with the advent of CAD and CNC became completely digitally made. This change resulted in enormous advantages from the quantity point of view but not always from the quality of the result. In the past, make the maquettes, was a task of experts, and as Ettore Sottsass says talking about Giovanni Sacchi, was a sort of completion of the design process.

> Fuori da ogni lode generica, la sua grande capacità va oltre il « fare » i modelli: è il capire gli oggetti che poi, lui, con i modelli racconta… Con Sacchi si va oltre il volume: lui fa sentire cosa succede veramente, tattilmente: produce una sensazione evoluta, tanto che un suo modello può soddisfare completamente il designer. Con un modello così, in verità, non si ha quasi più voglia di fare l'oggetto [13, pp. 119–123]
>
> «Out of all generic praise, his great ability goes beyond "doing" maquettes: is understanding the objects that then, he tells with maquettes …Sacchi's maquettes go over the volume: he makes you feel what happens really, tactilely: produces an evolved feeling, so much that his maquette can satisfy completely the designer. With a model like this, in truth, there is hardly any desire to do the object.»

Today many conditions have changed and consequently changed the design process, with the possibility of independently produce high quality maquettes.

2 Desktop Digital Fabrication

In this "*new industrial revolution*" the most significant change is not about technology but about the use that is made of this, those who use it and where does it. Some authors speak about technology's democratization, or better access to that. "*The atoms's revolution is bringing into our homes machines that can create real objects. These machines require a few tens of minutes to turn a three-dimensional model, a design made on a computer, in something physical, that we can touch.*" [10, p. 124].

We are talking about additive construction machines (such as 3D printing) and subtractive construction machines (such as milling cutters and laser cutting machines with numerical control) that can easily shape the starting material from a CAD drawing.

This change offer the opportunity for a critical consumer to self-produce his own design solutions, or as Anderson says: "*Mass production is good for the masses; but what works for you?*" [1, p. 76]. So, areas of special needs's consumers, often struggle to find on the market reasonably priced products that solve their specific needs. We think about the world of the blind and visually impaired, which often depend on the others sensitivity (those who design and build the living and public spaces) to easily interact with the world around us. With the aim of testing the potential of the Digital Desktop Fabrication for the users with specific needs, was created a co-design workshop with blind users. For the work's realization has been chosen a conventional modeling workshop, the carpentry of Archivio Giovanni Sacchi (Sesto San Giovanni, MI) (Fig. 2).

According to the ICSID (International Council of Industrial Design) "*design is a projectual process that aims to solve specific problems of society,*" an ambitious goal because the specific problems belong to individuals rather than to society. Indeed in the Design for All approach, solutions are designed in an inclusive manner, trying to solve the problems of the greater amount (and variety) of people, pushing to consider people with specific disabilities. According to Paul Hogan, president emeritus of European Institute for Design and Disability, (Design for All Europe) "*Good design enables, bad design disables*".

The best design, however, is individual and specific. In the case of the mentioned experience, visually impaired users, "*experts in their own problems*", have been helped by the designer in the project of their problems's solutions, which can be realized (for production's cost reasons) only thanks to the Desktop Digital Fabrication.

3 Physical Computing and Interactive Prototypes

Besides the possibility of producing in an automated way few pieces of a product, or a complete product, now has spread the need to animate or give "intelligence" to these products. The physical computing is the research's area interested in studying this relationship between physicality and behavior. The physical computing deals with creating a communication between the physical world and the virtual world of computers [12, p. xix], thus giving intelligence to products.

The industrial products can be considered intelligent when they are able to gather information from the context (using sensors), to process these information to make decisions according to a default logic and posteriorly act to change the context (using actuators). A clever product, in summary, is composed of sensors, processors and actuators and the behavior of this product depends on the decisions that the designer assigned to all environmental conditions.

Fig. 2 Detail of table design L'Edere e Maestroni, responsive morphologies course 1st Ed

In this way the firmware (the program that determines the behavior) becomes a constituent part of the product's construction information, as are the technical drawings, the indications on the materials, dimensions, manufacturing processes, etc. We can say that an intelligent product consists of software and hardware, of material parts (hardware) and intangible parts (software), and that the physical computing deals with the study of the connection between these two parts.

For example, about how a sensor (material) can gather information from the environment (material) and digitize it in order to be processed according to a certain logic, defined by the firmware (immaterial) which as a result will generate a return in the environment due to an actuator (material) (Fig. 3).

> Physical computing is an approach to learning how humans communicate through computers that starts by considering how humans express themselves physically. A lot of beginning computer interface design instruction takes the computer hardware for given - namely, that there is a keyboard, a screen, perhaps speakers, and a mouse - and concentrates on teaching the software necessary to design within those boundaries. In physical computing, we take the human body as a given, and attempt to design within the limits of its expression. [12] 1

This situation radically changes the process by which the products are designed, if in the past the representation was a sort of codified synthesis of the project's complexity, at this time, the representation is part of the project's definition and complexity.[1]

[1]For example, Aibo, the Sony's robotic "dog", can be upgraded with new features that take advantage of the best available hardware (sensors and actuators).

4 Computational Design

Since the manufacturing machines can interpret a digital document and produce a material object with few formal limitations and a very high precision (compared to analog machines) and that these machines have a simplified use even for non-experts, many designers have began experimenting with bits and computer science.

So, from a three-dimensional modeling that used to work with the industrial production paradigm (designed according to what could be mechanized) we are switching to a mathematical and, if we want, abstract modellation of the production's limits, with the knowledge that the digital production machines can produce "*almost everything*".

The accessibility to the formal experiments allowed by desktop machines, means that we can push the formal research up to limits never thought before or possible only to experts (Fig. 4).

As an example we can cite the design of a custom shin guards, developed by the Physical Computing Laboratory (PhyCoLab) of the Politecnico di Milan for a child.

The specific need was to cover from being hit but not from the air an injury resulting of surgery. A product that does not exist in the market and that would not make sense to produce industrially for cost's reasons in connection with the request's frequency.

The example given has been designed with an algorithmic software that allows to change the size and formal relations between the parties. This feature simplifies scaling and adaptation to the anatomy of the user, as well as the product's aesthetics, aware that there are almost no limits production thanks to 3D printing (Fig. 5).

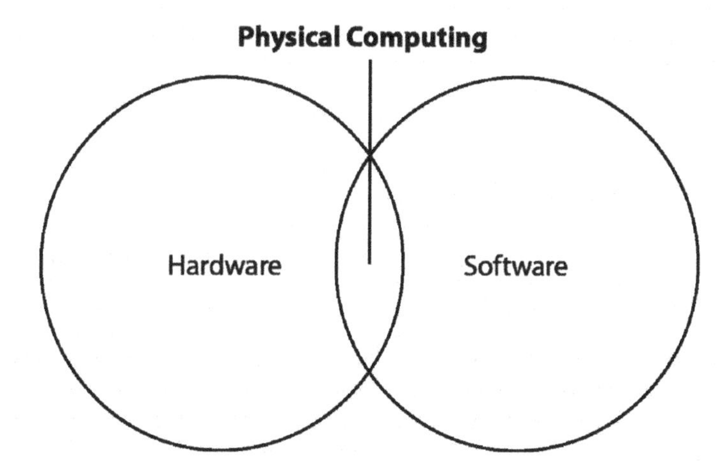

Fig. 3 Designers and blind users co-designing products to be printed in 3D

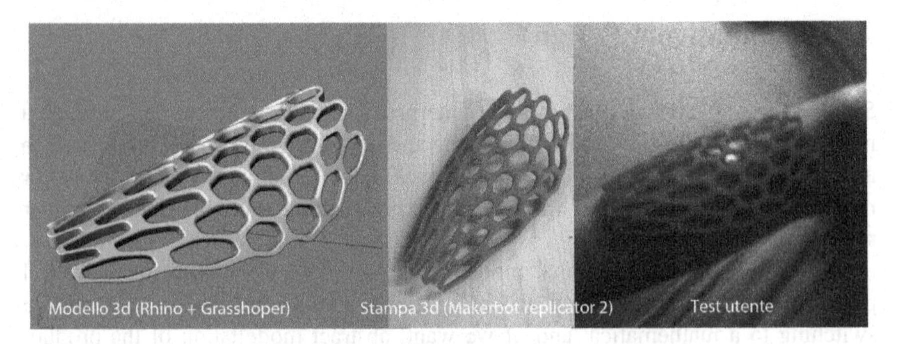

Fig. 4 Physical computer involves hardware and software

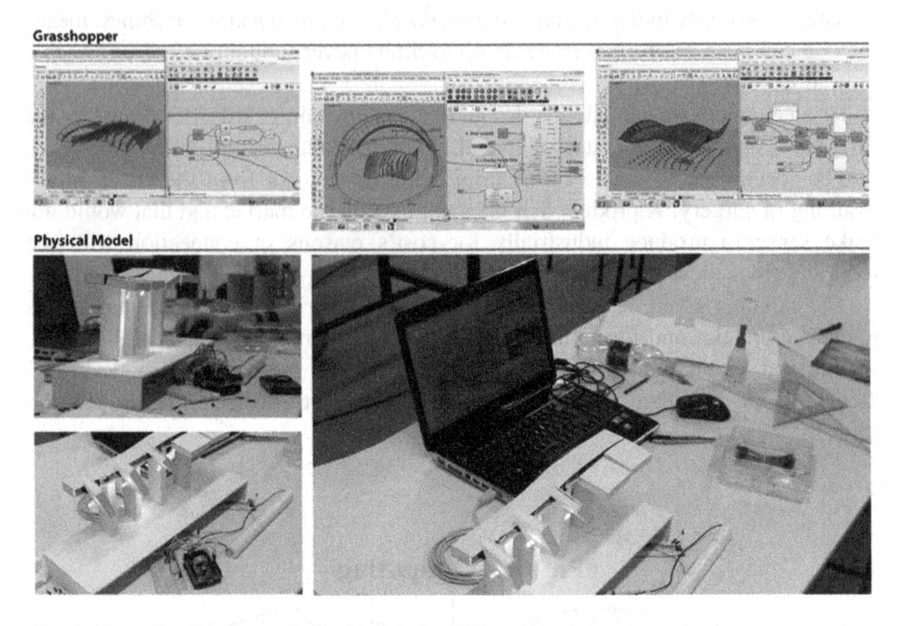

Fig. 5 Example of process of algorithm design, 3D prototyping and user testing

5 The New Design Paradigm

We can guess that the set of knowledge related to Computational Design, Physical Computing and Digital Fabrication are becoming the new paradigm of complex design.

The relationship between these disciplines is evident in project activities where the result to be obtained is represented by devices able to interact with the environment. The design process used by L'Edere and Maestroni for the design of a reactive canopy in the context of the responsive morphologies course 1st Ed. of Politecnico di Milano demonstrates how the iteration and repetition of the cycle design-prototyping is a fundamental condition in this context. In fact, in complex and interactive projects, the necessity of an approach Trial and Error [5] is still critical, but now becomes accessible even at high levels precision and low cost.

The case of the project mentioned is an example of how is possible to teach new designers the use of these technologies and a design approach strongly contaminated with sharing and multidisciplinarity.

6 FabLabs and Makers

During 2001 in the United States was born the first FabLab, at the Massachusetts Institute of Technology, from an idea of Professor Neil Gershenfeld, thanks to a grant from the National Science Foundation, and in collaboration with the Grassroots Invention Group and the Center for Bits and Atoms (CBA), part of the MIT Media Lab.

The FabLab are Fabrication Laboratory, one of the most popular types of space dedicated to "*do*" in common, small laboratories that offer personalized services and digital fabrication tools like 3D printers, laser cutters and CNC machines. Nowadays there are FabLabs worldwide, from the US to South Africa, from Afghanistan to India and New Zealand to Brazil.

Gershenfeld creates his laboratory as a support to the course "*How to make (almost) anything*" (MAS 863) MIT, a course that comes from the idea that if people can rule the construction technology, with this they can build solutions to their problems and that these production technology today's are digital and accessible.

> I realized that the winning solution for digital production is the personal production. This is not to manufacture what you can buy at Wall-Mart, but to manufacture what you can not buy there. [8]

7 A New Teaching Method

In these education contexts has spread the use of the term "*Thinkering*" to define a learning method based on experimentation without predefined purposes. Banzi defines it as "*what happens when you look for something that does not quite know how to do guided by whim, imagination and curiosity.*" [2] and that has origins in John Seely Brown article describing "thinkering" as "constructing/playing/wrestling with objects by appropriating, transforming and personalizing them for one's own learning and practice" [4].

This process, which is based on experimentation and not only on the knowledge transfer from an expert to a student, is connected to the relationship between manual and intellectual work, because intellectual work is a result of a manual experience that takes place before and not the opposite. You learn by manipulating, not observing.

The dichotomy between mental and manual labor is largely developed by Matthew Crawford in his book "*Shop Class as SoulCraft*". In his work, the author proposes this thesis by a historical analysis of the work separation and the resulting alienation of some intellectual tasks that become completely disconnected from the physical reality.

This situation is often evident in the projects developed by design students, accustomed to designing exclusively with computers thanks to CAD systems, that fail to predict errors sometimes macroscopic to a trained eye in practice. The material reality of the projects does not forgive the mistakes that instead you can hide in the photorealistic graphical representations.

> You can not become a musicians without learning to play a specific instrument, without subjecting your own fingers to the discipline of the keyboard or strings. The expressive power of the musician is based on an obedience that comes first; its ability to act is dependent on its continued submission. … He owes obedience to mechanics reality of his instrument. [6, p. 64]

It becomes clear that, in a world that is about to enter in a new industrial revolution, the designer's formation should be reconsidered.

8 Conclusion

More than twenty years after the industrial design review by Tomas Maldonado, the basic role of the designer is still in effect, although the production capacity and the user's needs have changed greatly:

> El diseñador industrial ha tenido siempre la función de proyectar productos industriales mediando entre las capacidades productivas y las necesidades del usuario. [11]

> The industrial designer has always had the function of designing industrial products mediating between the productive capacities and user's needs.

Already in 1993, in fact, this author had highlighted the problem of the role of an industrial designer who has to do with an area where microelectronics and computer science are pushing to the miniaturization of technology products.

> Muy a menudo, hoy en día, proyectar objetos no es de hecho muy distinto a proyectar sistemas de interacción. […] Sin embargo eso no impide que una parte considerable de la actividad del diseñador industrial siga firmemente anclada en la tarea de "dar forma" a objetos materiales que, nos guste o no nos guste, siguen estableciendo una relación tradicional con los usuarios, es decir una relación que se desarrolla, precisamente, a través de la naturaleza materialmente tangible de los objetos. [11]

> Very often today, designing objects is not actually very different from designing interaction system. […] However that does not prevent a considerable part of the industrial designer activity remain firmly anchored in the task of "shaping" physical objects that, whether we

like it or not, continue to set a traditional relationship with users, a relationship that is developed precisely through the physically tangible nature of the objects.

As anticipated by Maldonado, Industrial Design began to be increasingly faced with the need to draw "shells for printed circuitboards and electronic cards" without even understanding what there was inside, but knowing only that the user had to interact with some kind of display.

Thus, as suggested by Bonsiepe [3], some designers began to be interested in the design of the information contained in the interfaces of these devices, even if they were mostly the engineers to develop visual digital interfaces [7].

Today, the interaction man-product is much more than the graphical interface, the full product has communication skills, not only formal but also behavioral. Moreover, the product becomes the interface of a complex system and is connected through the internet to various services (Internet of things).

The Italian design is still on the edge of this change and could remain out. Currently on the international scene of human resources recruitment for the design, in addition to the typical figures as Product Designer, Designer Transportation, Car Designer, Interaction designers and others, began the search for profiles like Wireless Product Designer, Mobile Product Designer, Product Interactive Designer which together with Interface Designer occupy the design spaces for the new type of electronics basic products: ICT products. In this context the designers have the role to materialize in the best possible way the digital technology.

One of the key figures is defined as Design Technologists, who should help to overcome the gap between design disciplines and technological ones. The specialist who heads the creative team and leads to the implementation of design solutions ensuring technical capabilities and conformity with the requirements. A designer technologically evolved. The Design Technologist must have knowledge of Physical computing, Digital Fabrication and Computational Design.

By analyzing the results of the industrial design's great contests at international level as Red Dot, Good Design Award and the IF design, we notice the almost total absence of Italian names in sections related to ICT. If Italy has always had an important role for example in Car Design and Furniture Design, now in consumer technology products the latest innovative trends from the point of view of style, morphology and interaction, are coming from the east or from the United States of America. One reason for this absence could be attributed to the fact that in Italy there is still a shortage of ICT companies, and therefore the conditions for specific figures formation in this field have not been established. The Italian production system is based mainly on small and medium enterprises specialized in traditional manufacturing sectors, little technological. This meant that the designers in Italy have developed a great experience in the design of products such as furniture and furnishings in general. Therefore there are few designers who design household appliances and even less those who design products of consumer electronics or ICT. Although in recent years the Italian government, following the European lines, incentives the creation of high-tech companies to revive the national economy worldwide, it is evident that, until now, the specific preparation level in the field of ICT of the Italian designers is scarce.

The ICT is complex because it changes quickly: technologies evolve vertiginously and it is difficult for an historical academic structure to follow its progress. At this point it becomes important to have direct contact with the sector companies, which by their nature must even anticipate these changes. In order to form the new designer's profile is therefore necessary to push the creation of laboratories that allow simultaneously design, testing and evaluation of products with the end user.

References

1. Anderson, C. (2013). *Makers. Il ritorno dei produttori. Per una nuova rivoluzione industriale.* Milano: Rizzoli Etas.
2. Banzi, M., (2008). *Getting started with Arduino.* O'Really Media.
3. Bonsiepe, G. (1991). *Il progetto dell'interfaccia.* Línea gráfica, no. 3.
4. Brown, J. S., & Duguid, P. (2000). *The social life of information.* Cambridge MA: Harvard Business School Press.
5. Campbell, D. T., Stanley, J. C., & Gage, N. L. (1963). *Experimental and quasi-experimental designs for research.* Boston: Houghton Mifflin.
6. Crawford, M. (2009). *Il lavoro manuale come medicina dell'anima.* Milano: Mondadori.
7. Cooper, A. (1999). *Il disagio tecnologico.* Milano: Apogeo.
8. Gershenfeld, N. (2005). *Fab: The coming revolution on your desktop.* New York: Basic Books.
9. Gershenfeld, N. (2012). *How to make almost anything*, Foreign Affairs, Novembre/Dicembre.
10. Maietta, A., & Aliverti, P. (2013). *Il manuale del maker, La guida pratica e completa per diventare protagonisti della nuova rivoluzione industriale.* Milano: Edizioni FAG.
11. Maldonado, T. (1993). *El diseño industrial reconsiderado.* Barcelona: Editorial Gustavo Gilli.
12. O'Sullivan, D., & Igoe, T. (2004). *Physical computing, sensing and controlling physical world with computers.* Boston: Thomson Course Technology.
13. Polato, P. (1991). *Il modello nel design. La bottega di Giovanni Sacchi.* Milano: Hoepli.

Walkacross: Space–Motion Metric for Responsive Architecture

Paloma Gonzalez Rojas

Abstract Nowadays it is possible to collect more data than ever before. Yet, producing meaningful insight from such data is perhaps the most difficult task. This is particularly true when collecting data from people as for attempting to make architecture responsive to them, or for informing the design process. For example, some buildings respond to the environment temperature or light in real time. In such cases sensors trigger precise responses; high radiation equals to producing more shade; radiation is a straightforward indicator. However, determining how architecture might become responsive to people's spatial behavior is far from being a straightforward matter. Human–space interaction has been barely studied and usually is defined in not quantifiable terms. To address this quantification difficulty, I identified motion as the main indicator of human–space interaction. Motion is quantifiable and is composed by space and time. Motion is understood as a sign of people's perception of space. Furthermore, the goal of this study is the development of empirical research and data analysis of how architectural spatial configuration affects people's motion. The architecture design process is the focus to assess meaningful insight. A metric, denominated space-motion, was developed as a result of the need to categorize data. The space–motion metric is composed by 6 indicators that correlate people's motion with architecture features. The data collections for the analysis were recorded with Microsoft Kinect, tested in several data collections at the Massachusetts Institute of Technology and the world. Kinect data, assessed with space–motion metric, might bring unforeseen advancements for contributing to responsive architecture and architecture design process in general.

Keywords People's motion · Human–space interaction · Kinect · Metric · Data

P. Gonzalez Rojas (✉)
Design and Computation Group, Massachusetts Institute of Technology,
Cambridge, MA, USA
e-mail: palomagr@mit.edu

M. Rossi and G. Buratti (eds.), *Computational Morphologies*,
https://doi.org/10.1007/978-3-319-60919-5_12

1 Overview

This paper seeks to respond to the question of how to translate people's motion in space into quantifiable parameters. It also proposes developing a system of measurements to assess data that provides meaningful information for designing architecture. The hypothesis consist of: if we understand people's motion as a sequence of actions, then the actions could expose parameters or indicators that define them. Such parameters are correlated with architecture features with the aim of withholding the geometry that produced the motion (Fig. 1).

Responsive architecture frames this research; such type of architecture seeks to be responsive to humans, among other aspects of the environment, which is the proposed application of the research presented in this paper. The first section of this paper presents the work of designers and researchers that have studied how responsive architecture is defined and how people's motion in space is characterized in terms of design. As a disclaimer, this paper does not intend to theorize about responsive architecture or give an historical reference of it. Responsive architecture is included here as the framework for inscribing the study of people's motion in space, which is the focus of the research. The second part of this paper refers to motion as the main parameter to analyze people's behavior in space, and then presents the data collections and space–motion metric. The data collections are the foundation for the analysis of the effect of architectural features on people's motion. Based on the data analysis, I performed an iterative process to identify the indicators of the space–motion metric. Following, the metric was implemented to assess the data. The space–motion metric systematizes the process of analyzing spatial behavior of people to meaningfully inform design. The steps taken for the experiment: collect data, visualize the data, inspect the data for identifying the indicators for the metric, and development of the metric. Finally, I present the discussion and conclusions from the work and questions the main ideas of this paper. This paper is an extract of my Master's Thesis "Space and Motion: data based rules for public space pedestrian motion" developed at the Massachusetts Institute of Technology, with which I graduated in June 2015. The extract presented in this paper focuses on the development of the space-motion Metric.

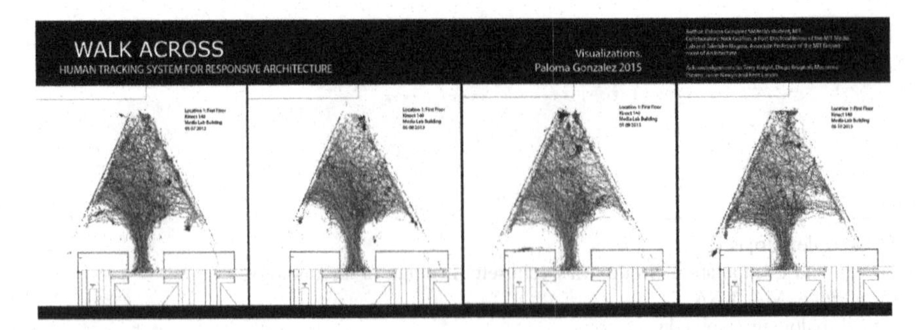

Fig. 1 MIT Classroom 140 Elevator Entrance Series

1.1 Framework

Responsive architecture is a relatively new approach to adaptable architecture that responds to the changing conditions from environment and users. In this paper I refer to responsive architecture in a practical manner, applying its concepts when analyzing people's motion in space. The research follows this hypothesis; if we can understand how people behave in space and predict their motion, we can then propose buildings that might "read" what its users are doing and adapt without them controlling the adaptation. In type of adaptable architecture the goal is that the buildings respond to its user by rearranging their form if necessary, as for achieving energy performance optimization and/or occupant comfort. Nicholas Negroponte coined this expression in the seventies, seeking to include architecture in the technological stream of embedding sensors and actuators into buildings for turning these constructions into smart objects [1]. In 1970 Nicholas Negroponte and the Architecture Machine Group (MIT) developed an early and famous example of a prototypical responsive environment for the art installation "Seek," included in "Software" exhibition at the Jewish Museum in New York (Godinez, n.d.). The project consisted of a small environment for a gerbil colony to inhabit, composed by metal blocks inside of a glass box. The gerbils were continuously observed in real time to identify motion probabilities inside the environment, and a robotic arm rearranged the configuration of the blocks according to the gerbil's most likely future actions. Even though this experiment might be considered a fail because the environment did not adapt to the gerbil's behavior, it encompasses the basic concepts of responsive architecture. The piece exposes the idea of transformable and adaptative environments that shapeshift to accommodate to changing conditions. More importantly, the piece gives clues about the need of quantifying the behavior of the occupants of a space and the search for reliable rules to adapt the environment properly. Perhaps, the miscue was that analogue observation was a deficient input for deciding how to rearrange the environment; a tracking system might have been an improvement. In addition, probabilities were set a deterministic, "programmed to either correct or amplify (not both) the dislocations caused by the gerbils." (Godinez, n.d.). Currently the "Changing Places" group of the MIT Media Lab is working on the "CityHome" project, which one might say is a human-focused and more sofisticated version of this environment for humans. The project is a small apartment of 78 m^2, with a robotic wall that changes its form. The size constraint justifies the need of spatial transformations as the wall shape-shifts according to the program. In this case conscious gesture commands trigger transformations of the furniture. In the previous examples, the environment adaptation towards users was conducted dynamically in real time, presenting some of the most significant works of responsive achitecture.

1.2 Related Work

Up until current times, a modest number of architects have attempted to analyze people's behavior in relation to designed space. Chong et al. [2] research is a great

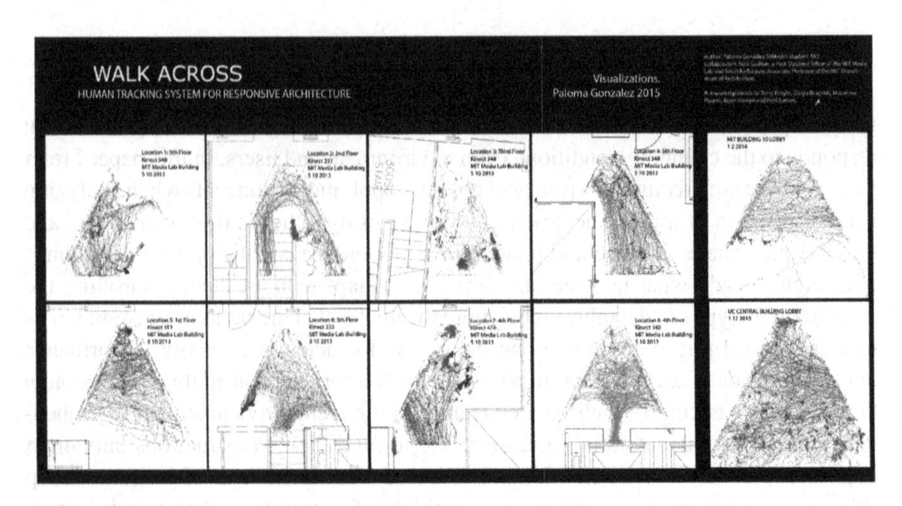

Fig. 2 MIT Media Lab Visualizations

precedent for such regard; their work devolves by observing people in space and asking about their experience in each of the settings. The study describes static human actions in space, as people were asked to stand at some position of a room and observe from there. The study contributes with a starting point for the analysis of human spatial behavior that leads to the following stage of analyzing motion (Fig. 2).

Cristián Valdés, a Chilean architect from the University of Valparaiso, measured and took notes of how a family occupied a house for a month. Cristián Valdés registered their motion trajectories to understand space, as part of his final undergrad project at the Catholic University of Valparaiso [3]. In a series of plans, he marked dots where the people of the house spent time during the day. By the end of a month he obtained a blob showing the traces of people's motion in space. This architect defines the use of a space as following a "law" and the law was represented in the final drawing depicting the blobs. This example is makes explicit two aspects of human-space interaction: first the difficulty in representing motion in static means such as paper drawings. Second, space and people's motion is a matter that has consistently intrigued designers.

The research process developed by Cristián Valdés depicts how people's motion could inform the architecture design process [4]. It is clear that motion should be complemented with other design variables to deploy a consistent argument for an architectural project, which is a shared opinion in the context this research. However, people's motion has the potential of informing design and creating a different perspective for creating space.

The urbanist William H. Whyte, in "The Social Life of Small Urban Space" [5] developed statistics from small urban spaces. Whyte and his team video recorded public spaces filled with people and analyzed the footage. The statistics where mainly focused on the activities performed by people in relation to elements of the square such as trees, shadow areas and benches. Yet the study accounts the activity more than the dimensions. Taro Narahara presented a similar research for his Master's Thesis "The space Reactor" [6]. Narahara's research proposes to measure motion against volumetric features of space. Narahara developed a simulation and analyzed the simulated behavior of people. The research is focused in the "reactions" of people towards spatial elements.

A further important related work is Space Syntax, a software and a theoretical model developed by Bill Hillier and Alasdair Turner ("Space Syntax | science-based," n.d.). Such model seeks to provide understanding about people's behavior in space by investigating for people's perception of the configuration of space. For example, Space Syntax model is used to describe space a network of nodes, disregarding the three dimensional geometry of the elements that compose space. In a sense, this definition of space delineates spatial topology. The model proposed in this paper seeks to analyze people's motion in three dimensional space, to study people's motion and how architecture features affect that motion. Understanding architectural features are defined as their geometry or volumetric description and position in space, not as nodes.

Finally, [Seera et al. 7] proposed using Kinects to validate analyzing people's flows in space, by installing the sensors at the MIT infinite corridor. The researchers have contrasted these data with the flows produced by simulation software. This paper is a great reference for the use of Kinect sensor as a reliable tool to track people's motion. However, the purpose is radically different from the one proposed in this paper.

2 Method

The problem that this research explores is developing the means to design space through motion. This line of questioning alludes to the prevailing conception in architecture of designing the static form of space, which is tangible and possible to grasp, over the changing dimension that holds problem, as Michelis [8] and Latour [9] confirm. This changing dimension, often explored in architecture as light transformations, is understood here as transformations in people's motion. By incorporating motion into the design process, time as the fourth dimension is introduced as well, as motion entails to consider the project in terms of a process, or more specifically, as a sequence of motion.

Defining why people behave in a certain manner seems to be a boundless problem; from personal desires to random conditions imposed by the environment.

Human motion sequences are translated into a set of actions, "how" a person moves in space, the trajectory the person takes, the bodily motion of the person, the rhythm of their walk. The paper's claim consists of: by recording and analyzing people's motion it is possible to extract indicators that affect people's motion in space. Therefore, this research maps, quantifies, and formulates pedestrian motion correlation with space and questions the role of data for projecting what space could be.

2.1 Data Collections

When humans move inside a space they leave an unseen trace that has not been possible to accurately record and reproduce until very recently. In 2010, a gaming depth camera, the Kinect, was first released, and in 2011 Microsoft released the SDK (Software Development Kit) generating interest in tracking people's moves and walking-paths. The Kinect tracks people's skeletons inside a space. Since Kinect are depth cameras it is possible to get the three dimensional model of the space and of the human. The device works as sonar, emitting a pattern of infrared light.

The original purpose of the Kinect was to replace the gaming controller for the XBox One, and therefore the platform for recognizing humans is remarkably stable. The Kinect recognizes up to six humans and can track in detail two of them. Nonetheless, it retrieves the trajectories of six people's center of mass at the same time. The human recognition algorithm is a major difference with video cameras in which the recognition is made through background subtraction. This makes Kinect especially appealing for tracking human trajectories. The sensing range of the Kinect is a triangle of 3.5 m depth and 3.0 m wide retrieving the position data of skeletons and video recording located in that range. The advantages of using Kinects over survey methods are significant; the data collection is highly reliable as the trajectories correspond with real time streaming, and the location data is precise to the millimeter.

Phase 1 Data Collection: consists of defining a data collection location, determining an advantage position for the devices, recording with Kinect and video camera, and observing people's motion on site. The data collections were performed in several locations, defining two of them as the main spaces to be analyzed. The difference with the rest of the sampling is that in those two spaces video camera was used, complementing Kinect sensor. In the rest of the locations the data collections were performed only with Kinect.

Phase 2 Data Analysis: consists of developing data visualizations, observing the frequency of motion, and developing statistics using the proposed metric. All the processes were performed with several algorithms developed in the context of this research. The data visualizations were developed using Processing Programming Language. The data visualizations function is to provide a platform to inspect the Kinect data and observe the frequency of motion. Through the data visualizations and the data statistics with Space and Motion Metrics, behavior patterns can be observed in time. Finally, the measured parameters are summarized in averages, percentages and other statistical indicators.

2.2 Data Processing

In order to study the datasets, I developed trajectory data visualizations, contextualizing the movement to the original locations. Eight of these data visualizations are presented in this paper, developed with the datasets from the Media Lab Building plus two samples from MIT Building 10's Lobby and the dataset from UC Central Building.

The visualizations show the paths of people recorded in an eight hours period. The data retrieves the coordinates and user ID. The color is changed every time the Kinect tracks a new user. The locations are diverse; ranging from a first floor entrance, to a staircase with slow pace. It is interesting to observe several days of the same visualization since they show consistent patterns of behavior, as we can see in the "Classroom 140 Elevator Entrance" Series. The pattern of walking in a certain distance of the elevator is repeated, without having any element that forces the minimum distance to the door wall. Perhaps the proportion of the waiting area promotes the action of standing and walking to the center of the free space.

3 Results

How people interact with architecture is a considerable complex phenomenon and even though the Kinects retrieve 3D skeleton information, I developed a code that focuses on plan trajectories in order to constrain the analytical process. The workflow consists of developing the capturing code using Processing Simple OpenNi Library. This is a relatively simple process since this library has most of the basic methods of recognizing a user and translating it into a skeleton of 25 joints with xyz coordinates. The visualizations of the data have a triangular shape making explicit the sensing range of the Kinect and are overlapped with a plan of the location where the data was collected. Different depth sensors from the Media Lab Building data show different trajectory patterns according to each spatial configuration, such as people avoiding a wall corner or entering elevator doors. A noticeable difference is found between the Media Lab data and the data from the UC Central Building Lobby: in the first case the trajectories are straightforward, almost like lines in a functional path, and at the event the lines show many nodes, depicting the loose mood of the situation. In the case of "Lobby 10" visualization, the highest speed of the trajectories recorded from the MIT students was 2.4 m/s, much faster than the average speed of people, which is 1.4 m/s. Speed of the trajectories characterizes the universe of recorded people as in this case the students are between a certain age and similar life style. The Media Lab user maximum speed was 1.0 m/s and UC Central 0.9 m/s.

In addition, I could observe some spatial features that produce a common behavior such as the importance of the large voids that connect several stories of the Media Lab Building, as we can see in the "Classroom 348" and "Classroom 474" figures. In both of these visualizations large stains appear reflecting longer periods of time spent in the areas that are closer to the voids banisters.

This discovery may seem obvious,—the building is working as the architect's intention set it to be. Nevertheless, it was not until recently that this could be finally proven with field work data, which constitute the main contribution of this paper. The procedure of testing Kinect's capabilities for analyzing people's motions was developed through several museum installations. I have authored two exhibitions named "Walk Across" as the result of an independent research process for Prof. Takehiko Nagakura; one at The MIT Museum and one at the Harvard Graduate School of Design Kirkland Gallery. The first exhibition at the MIT Museum was exhibited to the public on May 2014, until the present, and the second at Kirkland Gallery from November 22th to November 26th of 2014. The data visualizations that I developed from Media Lab Building Kinect data inspired "Walk Across." The interactive installation consists of integrating Kinect sensor with a custom algorithm that displays real time visitor trajectory data in aerial view. "Walk Across" encompasses the concepts of my ongoing work; from the memory of space displayed in the visualization to the awareness of the body while moving; from making explicit the interaction between people, between present and past data, and with space, to collectively drawing with your own body. Finally, from seeing motion and seeing the disappearing of the re-creation of motion functions similarly to a simulation. Resulting in profound insight for my research, the work proved to be valuable even outside the discipline (Fig. 3).

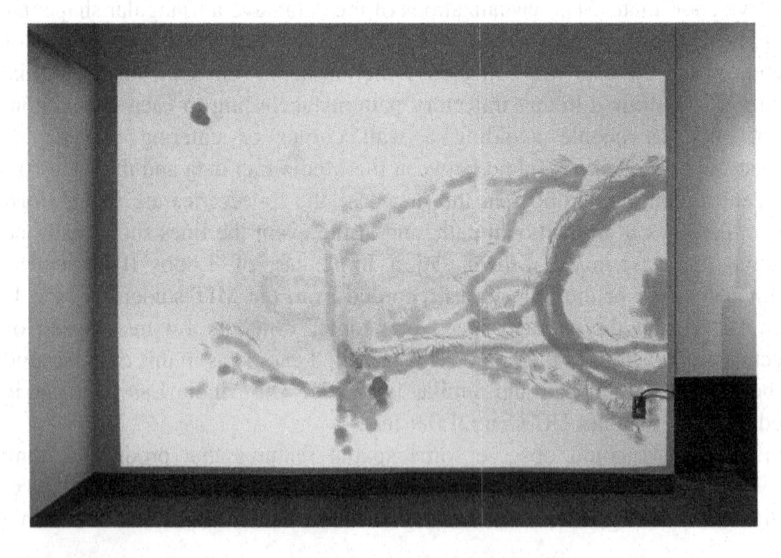

Fig. 3 WalkAcross MIT Museum Installation, credits David Schalliol, 2015

3.1 Motion Metrics: Parameters Regarding Space

The proposed space-motion metric consist of: 1-Speed The main question to answer regarding speed and space is: when does the speed changes in relation to spatial configuration. Speed exposes different states of people's motion, the period when a person starts walking or approaches an objective are slower, for example. Speed may also expose the effort that a person does while moving over a terrain. An interesting evaluation about speed is to observe if people change the speed value while walking from one spatial configuration to another, such as from walking from a wide space to a narrow space (Fig. 4).

Speed is measured in the Kinect range. Speed exposes different states of people's motion, as for example, to make explicit people's gestures when approaching a target or obstacles and the interaction with others. The baseline speed is 1.4 m/s referenced from the research developed by Mohler and Thompson [10]. At 1.4 speed meters per second a person is comfortably walking. In this study, is defined that a person starts to move from a speed of 1.0 m/s. The indicator for standing people is defined as speed of less than 0.8 m/s (Fig. 5).

2-Gestures. Gestures make explicit people's interaction with space. People perform a variety of gestures while crossing a space that may not be related to the space, therefore evaluating spatial gestures is a task that presents a challenge. After observing people's behavior for extended periods of time it is evident that the head and legs are the body components that participate in the interaction with space,

Fig. 4 WalkAcross MIT Museum Installation, credits David Schalliol, 2015

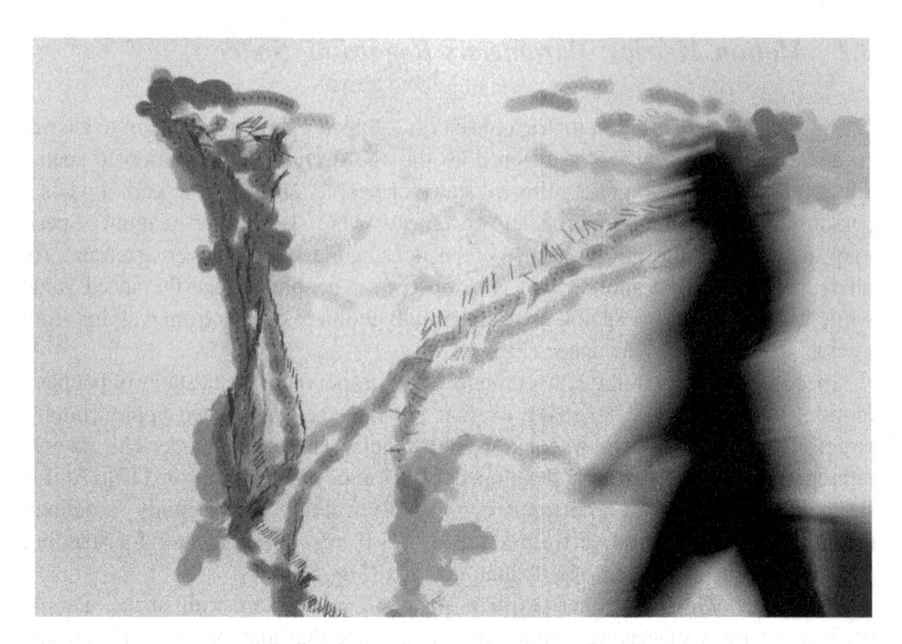

Fig. 5 WalkAcross MIT Museum Installation, credits David Schalliol, 2015

while in motion. According to this the gesture recognition is defined as identifying bodily movements from the head or the legs in interaction with spatial features. Gestures regarding space, are measured in video footage by manual counting. The gesture is composed by a sum of movements such as the change of the direction head, plus a change in the speed, for example. These gestures are defined by bodily motion regarding an spatial feature, and can be recognized in further research by the sensor through skeletal based recognition, currently built in the sensor Kinect.

3-Time. The data retrieve traffic flows in time, showing periodicity of the space, following Cristian Valdes words, the rhythms of space (2015). If people's motion is recorded for a long period, the data can retrieve how much time a person stays in one position and where people spend more time. With this information designers would know how to coordinate different events in a space. Timing people's motion in space can also expose the areas where materials suffer intense use. More importantly, timing space introduces periodicity to the conceptualization of architecture. This indicator is useful for defining the time a simulated person would spend in a certain spatial setting. In addition, with this information designers would know how to optimize a space or the areas of the space.

4-Shape. The shape of the path in plan could inform the route that people decide to take in a specific space configuration, and the characterization of the motion, such as caution motion, determined motion or wondering motion. For example, in the case of the elevator at the MIT Media Lab Building in the first floor, people take a curve to avoid the exterior corner of the elevator. The data collection retrieves the exact path, exposing the curve, exposing a caution motion.

5-Direction of the path. The direction of the path is defined by the sequence of points. Therefore, there is no need for origin-destinations surveys. The direction retrieves information about where the person is heading. The data that is taken into account in this research is three dimensional, and the direction is interpreted by analyzing the possible heading towards a spatial feature. The direction of the path could also be used to identify how many users have the same destination.

6-Scale refers to size of the analyzed spatial situation in relation to human scale. For example, the scale of an entrance is considered a micro scale situation; a public space of a hypothetical size of 100 by 100 meters would be large scale situation.

In addition to previous motion metrics, macro-scale cultural elements are embedded in each particular locations' population, and that motion is conditioned by cultural factors. Therefore, a second metric might incorporate the cultural associated spatial behavior to make explicit the location where the data comes from. In a sense, such metric is defined as an ethnographic measurement. That metric may also include the cultural aspects of the location regarding time. For example, for some cultures it is uncommon to walk extremely fast. Such characteristic impossibilities transposing the data to a space that is located in the context of a culture with a different speed range. Hence, it is essential to assess motion in terms of the cultural model from which the data comes from. Therefore, this type of analysis must be generated taking into account that it might be applicable only for the specific context it was produced. Architecture seems often linked to the place in which it is built.

4 Discussion

Robin Evans's research in "Figures Doors and Passages" explains how the invention of the corridor is a clear example in which an architectural element is used to modify human behavior [11]. That discovery was made by studying how humans inhabited their houses in the past, through history and documentation, without observing or tracking humans. Currently is it possible to track people's motion a myriad of devices such as video camera, depth sensor, and global positioning system via satellite.

The possibility to store large amounts of data is also possible, and the interest to explore such data for several purposes such as design and urban planning is increasing. Yet, how architecture features are incorporated to understand the impact on people's motion?

My analysis demonstrated the need of quantifying how people interact with architecture, for incorporating it into the design process and in real time for the development of architecture's responsive adaptation. We as humans shape our buildings, and after, buildings shape how humans behave in them.

This research mayor contribution is to provide evidence of people's behavior in space, enabled by Microsoft Kinect technology. The Kinect is one of the most reliable human tracking sensor with which is possible to accuratelly track people's trajectories, surprisingly effortless with a very robust system. The only disadvantage is that the sensor's range is considerable small, approximately 9 m^2. The disadvantage of the sensing range is resolved in the installation "WalkAcross" at the MIT Museum by using multiple Kinects for generating the expected data set.

By correlating the data from trajectories with the architecture of the selected place, this research retrieved reliable data analysis regarding an architectural setting which may prove or deny architecture designer's assumptions over space. Being able to get feedback on spatial relationships based on data collection methods is a useful metric for designers who have no other tools to address these concerns besides their imaginations.

The development of the space–motion Metrics contributes to the analysis of people's spatial behavior. The definition of the six motion indicators: speed, time, gesture, shape, direction and scale are useful metrics to enable such empirical research. The development of the indicators of the metric required a considerable amount of data analysis, and observation on site. These indicators assess the interactions between people's motion and space, with the "spatial gesture" as the main conceptual development.

The metric was tested in all the locations. Speed is the most basic computation regarding position and time, yet, when correlated with a spatial feature acquires a different meaning. The speed that people show at free walking in a corridor seems to be affected by the configuration of the space. One could speculate that the width of the corridor can define the activities that are possible to do; a wider corridor allows standing in groups and a narrower corridor acquires the category of transit space, as a result of applying this metric. The research proposes to analyze the "spatial features" of a location, understood as the volumetric elements that compose an architectural setting. The Space–Motion Metrics measure "how" people interact with that feature, as modifying their speed, or changing their bodily posture. The definition of the spatial features considers a corridor as spatial element, or an entire space depending on the scale of the analysis. The analysis is builds upon the analysis developed by Narahara [6] and Whyte [5].

In terms of tracking people in real time for the dynamic response of a building to enable architecture to be responsive, this research is a small step towards enaling buildings to be responsive to people's motion. It is my belief that to be able to generate effective responsive architecture that responds to human-space interaction parameters it is needed to identify signs or gestures to respond to. Similar to Negroponte's experiment with the gerbil's responsive environment in which the behavior of the animals was interpreted to trigger the environment adaptation, the space-motion metric retrieves information for determining when to trigger physical adaptation for buildings to dynamically respond to humans.

References

1. D'estrée Sterk, T. (2003). *Building upon Negroponte: A hybridized model of control suitable for a responsive architecture*. Chicago: The School of the Art Institute of Chicago.
2. Chŏng, Y., Branzell, A., & Chalmers Tekniska Högskola. (1995). *Visualising the invisible: Field of perceptual forces around and between objects*. Göteborg: Design Methods, Department of Building Design, School of Architecture Chalmers University of Technology.
3. Iturriaga del Campo, S., & Valdés, C. (2008). *Cristián Valdés: la medida de la arquitectura*. Santiago de Chile: Ediciones ARQ : Escuela de Arquitectura, Pontificia Universidad Católica de Chile.
4. Gonzalez Rojas, P. (2015). *Space and motion: Data based rules of public space pedestrian motion*. Massachusetts Institute of Technology, and Department of Architecture.
5. Whyte, W. H. (1980). *The social life of small urban spaces*. New York: Project for Public Spaces Inc.
6. Narahara, T. (2007). The Space Re-Actor : Walking a synthetic man through architectural space (Thesis). Massachusetts Institute of Technology. Retrieved from http://dspace.mit.edu/handle/1721.1/39255
7. Seer, S., Brändle, N., & Ratti, C. (2011). *Kinects and human kinetics: A new approach for studying crowd behavior*. arXiv:1210.2838 [physics](October 10, 2012). http://arxiv.org/abs/1210.2838.
8. Michelis, P. A. (1949). Space-time and contemporary architecture. *The Journal of Aesthetics and Art Criticism, 8*(2), 71.
9. Latour, B. (2005). *Reassembling the social: An introduction to actor-network-theory*. Oxford, New York: Oxford University Press.
10. Mohler, B. J., & Thompson, W. B. (2007). Visual flow influences gait transition speed and preferred walking speed. *Experimental Brain Research, 181*(2), 221–228. doi:10.1007/s00221-007-0917-0.
11. Evans, R. (1997). *Translations from drawing to building* (p. c1997). Cambridge, MA: MIT Press.
12. Godinez, R. (2015). *Nicholas Negroponte with the architecture machine* (n.d.). Retrieved February 24, from http://rudygodinez.tumblr.com/post/78209795322/nicholas-negroponte-with-the-architecture-machine-group.

The Management of Parameters for the Design of Responsive Map

Cecilia Maria Bolognesi

Abstract The research deals with the management of data as parameters for the knowledge and design of towns. As environmental data we mean what the human being can offer in terms of habits, different age and sex, race and many others within the architecture of a town as a software in a hardware. The desire to represent social data related to human behavior of the inhabitants is an ancient wish. The paper is a short overview on the evolution of the collections of data and its representation in maps more and more complex. Nowadays territorial data bases are more often linked to time variables and need complex platforms to be described. Today complex software allow us to manage geolocalized datasheet where you can create synthetic but complex maps including different sort of dates. These maps can give us a direction of the state of the town but also a direction for a smart design and development. This papers show us an ancient case, some contemporary researches, an actual case in Milano.

Keywords Responsive map · Urban design · Real time informations

1 Social Data in the Map: A First Static Map

The map of London of Charles Booth [1] produces, after a long search, one of the first infographics of the century mapping a social data. It is not the case of a generative map or the process for a new design but the show of a social and environmental analysis. It is the poverty map in London printed between 1889 and 1891, in two volumes whose title was Life and Labour of the People. The issue of poverty in the cities under the influence of environment of the Victorian times was often discussed in the press: Booth recognized the importance of a correct description of people and facts related to a social view. From 1886 till 1903 he organizes a census [2],

C.M. Bolognesi (✉)
Department of Architecture, Built Environment and Construction Engineering,
Politecnico di Milano, Milan, Italy
e-mail: cecilia.bolognesi@polimi.it

M. Rossi and G. Buratti (eds.), *Computational Morphologies*,
https://doi.org/10.1007/978-3-319-60919-5_13

door to door, to collect informations. The object of the census are: London inhabitants, their social conditions, their job condition related to their salary and even if in a traditional way all data are georeferred to the map (Fig. 1).

Booth was strained by the complexity of the issue and by work without any real argument to support the target of his research. He tried in the colorimetric description a summary, something that could really take into account all the nuances that he actually met during his study. Social classes in which Booth divides citizenship range from A to F. In its definitions, we read: A, the more modest class, occasional workers, criminals with only luxury drink; B salaried occasional, very poor, workers up to 3 days a week; C artisans poor, crushed by the competition of the market, unloaders; D salaried regular but little such doormen, bellhops; E salaried regular, fairly well-off; F high class of workers, craftsmen paid workers with responsibilities; G lower middle class; H middle class higher. In the maps, yellow streets indicate the presence of upper middle class or the upper classes, the red and pink are synonymous with middle-class existence, or at least comfortable; then, it veers towards cooler hues, from violet, to the mixed conditions, the blue for the poor, to black, to the extreme poverty, coupled with vice and crime.

It can be difficult to interpret the colors used in the descriptive map of poverty in London. The choice was oriented in such a manner that colors of neighboring social classes were chosen not to emphasize large distinctions: in this manner adjacent classes have similar colors and the visualization concerns a more equal society. The result is an interpretation of the city with no major peaks but where specific identification at the street level is a bit 'more complex'. In recent times it proceeded to the digitization of all the materials for a subsequent georefering. With his

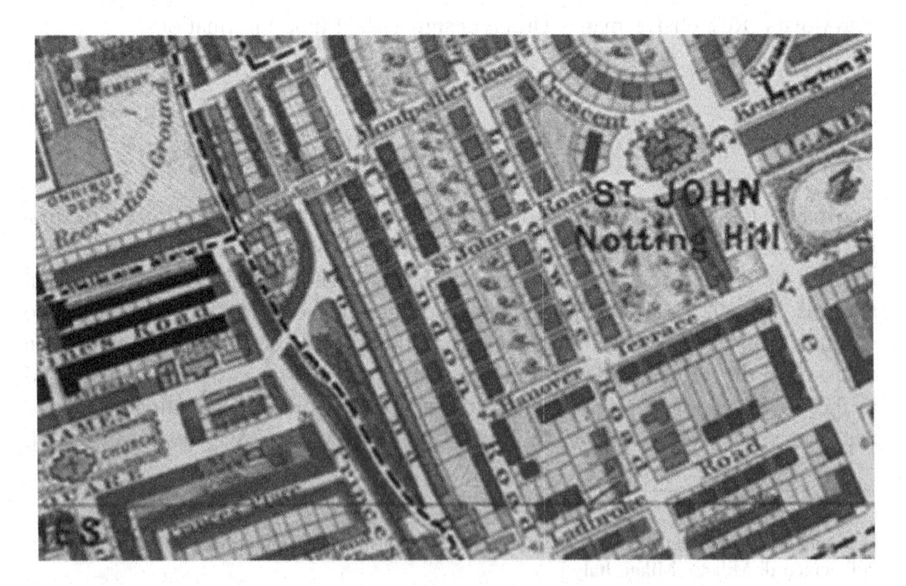

Fig. 1 Poverty maps of Charles Booth in London: lots were colored according to the social class of tenants, from the lowest and poor to the richest

colorimetry the map is one of the most famous case of a concrete visual link between the morphology of a town and his human capital, showing datas possible future strategies for environment. In fact, even if in a criticized mode it was used by administrators and urban planning of the city of London in the 60 for demolition of entire urban parts and tissue replacement. Over time the map turned out to be a data base of a series of data with smart original purposes cognitive with high social content and ready for immediate use in the context of urban development.

2 Time in the Map: Recent Responsive Maps

Isochronous maps add the variable of time within the geography of places. They represent travel times that differ depending on used transport that therefore generate different isochronous on the same routes. It follows a geography very different from the real one; It is however a real representation, tied to the speed.

The first experiments in this pertain to planning studies to large scale in Europe: the public transport that shortens the distance raises new issues in terms of planning. The objectives of these project are generally further development of theories on the interaction between transport and spatial development, improvement of planning methods and simulation and evaluation techniques and the identification of planning measures [3].

A very interesting case is provided by the European rail transport map drawn by Spiekermann and Wegener (Fig. 2). It is a forecast of the proximity between the cities based on the rail journey times in a perspective of development of the transport system from 1993. Their theory explains that modern transport technologies reduce the time to overcome space using spatial analysis and simulation, geographic information systems and advanced visualisation techniques; measured in units of time, space is ‘shrinking’ [4, 5]. Time–space maps represent this interaction between space and time cartographically described. In time–space maps the distance between two points is not proportional to their physical distance (as in physical maps) but proportional to the travel time between them. This change of map scale leads to distortions of the map compared with ‘familiar’ physical maps [6].

3 Other Data in the Map: Anamorphosis Maps

The previous cases maps the time factor connecting to distance where morphology is actually edited to match lengths and surfaces.

But there are many others interesting indicators that could enter in the discussion showing their connections with the map and modifying it in different ways.

This field refers to some maps called “*Anamorphosis Maps*” that we can consider responsive maps in the way that they modify their morphology in relation to

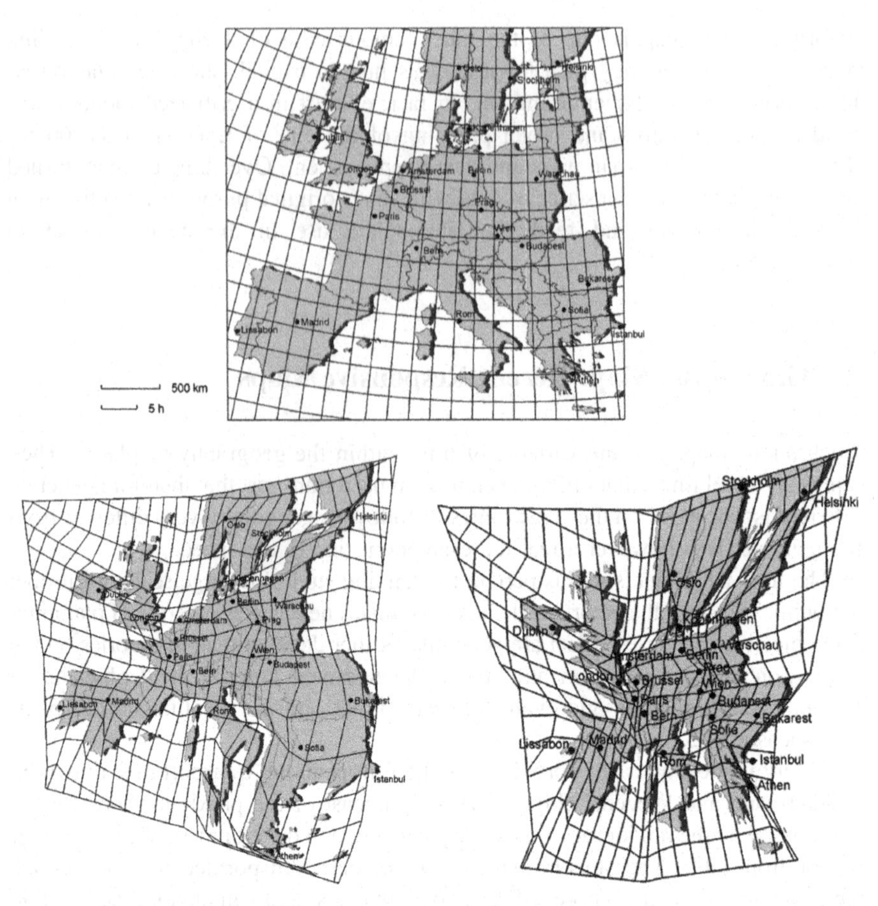

Fig. 2 The three maps show the 'shrinking' of space through the improvement of the European railway network between 1993 and 2020 compared with the base map with constant speed of 60 km/h. Spiekermann and Wegener Urban and Regional Research

the data (Fig. 3) making the original subject recognizable only through the interpretation of it.

In the beginning of the 20th century scientists have made efforts to plot economics material avoiding regular cartographic projections. German cartographer Vichel was the first who tried to use a distorted drawilig of a globe (1903). He published his maps of the "*Number of the population.*" Later this method was used for 'illustrations in its different modifications. New projections, which gave an opportunity to compose exact and strict anamorphosis maps, were then created [7]. The particular feature of these projections is in the inclusion of a main exponent in the equation and the opportunity to increase the amount of the information in the most compact parts of the map. The cartographic netting changes and the border of the continents of the states transform. It is still a problem to show a link of two

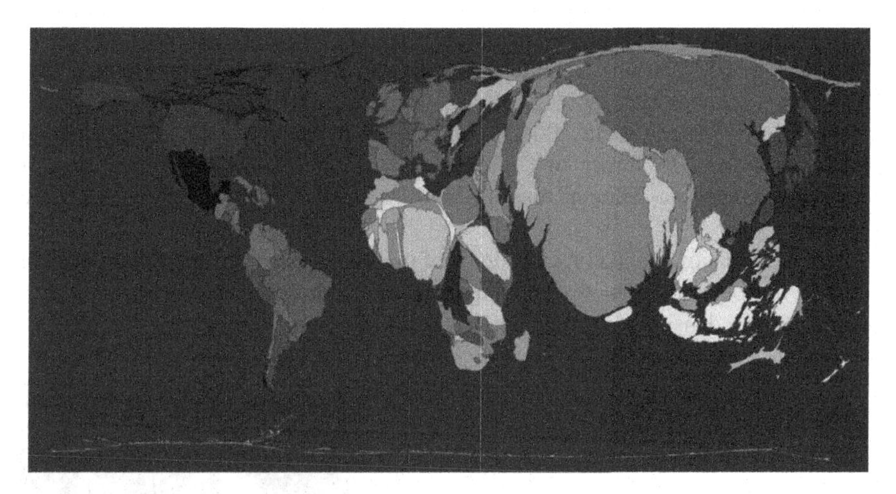

Fig. 3 A typical anamorphosis map on density of population in the world

different indexes on an economic map. The scientists continue the research works on the problem of creation the anamorphosis maps with the help of a computer.

One of the most famous is population density in the world where the deformation of the nation is the result of population density.

We are in front of changing map, collecting data, sensing data, acting in the plan. It is an unusual picture of the world. Where one can even not recognize the states on a very carefully depicted computer map, and it is nearly impossible to recognize them on a hand-made map. But this method allows to show expressively a real geographic situation. It gives an opportunity to compare economic or any other potential of different countries. It helps to make individual conclusions as well as to form personal opinion towards economic, social and cultural levels of a region or of a whole state.

4 Sensing Data, Actuating Instant Maps

At the beginning Booth worked a real census, door to door.

Human capital becomes part of the morphological data with a collection suited to the times.

The real story of responsive maps of cities, finds its moment of propaganda and dissemination among scholars in the Venice Biennale in 2006 with director Richard Burdett of the London economic School in London. In the title "*city architecture and society*" the director brings attention to the mapping of intelligent themes (smart) to give a better planning.

On that occasion Burdett presented some three-dimensional maps, real models that give shape to the population density of the world metropolises (Fig. 4). It is a new anamorphosis map where maps are sculptures at a scale that displays the density in square kilometers of high impact.

But still it is one data. We want here to remember what is experienced today in MIT labs, perhaps the first institute that started to deal with built environment with new disposal to collect data (Fig. 5).

In the same Biennale in 2006 [8], not far from the maquette of the London Economic School MIT map of Carlo Ratti maps data played by the impulses of smartphones using a different map reading the impulse and connecting it to time laps.

The information mapped regard flows of messages and call rings among Roman citizens in Rome during word cup final linked to monumental areas in the town. It is the show of a density of communication where a pointer shoes.

Fig. 4 A room of Venice Biennale of Architecture 2006_Density maps

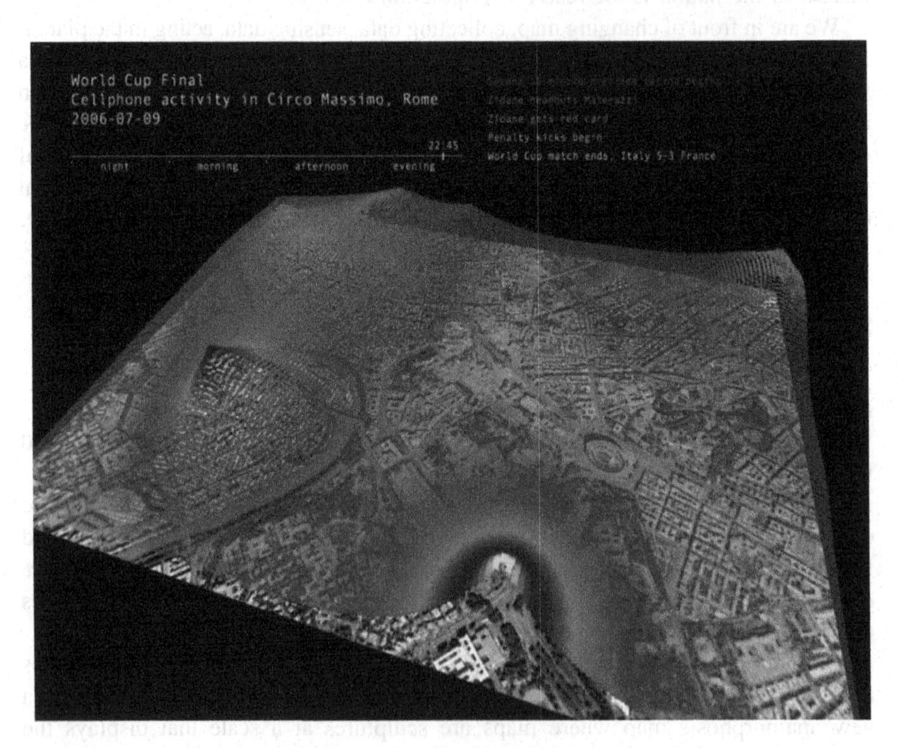

Fig. 5 Cellphon activity during Word cup final in Rome 2006

Smartphones, credit cards, GPS come in description of behavioral models of inhabitant of towns entering the world of the planning for the first time supporting theories that promote the model of Smart Cities.

In the theories of the Senseable City Lab, decoding electronic entity of the inhabitants of a town means recording behaviors and lifestyles. On the opposite some projects invert reading building with the data collected by the sensors developing environment of pure imagination. Decoding data can change behaviors and urban structures, cities in particular, according to a responsive model of sensing/actuating.

5 A Milanese Case History

In 2006 the Italian Builders Association [9], feeling the possibility of an incoming real estate crisis, decided to explore the possibility to have a scientific knowledge, a scientific check, on the territory where its members used to work.

That meant to think how to build a data base GIS located first of all on the territory of Milano, and possibly later, in its surroundings (Fig. 6).

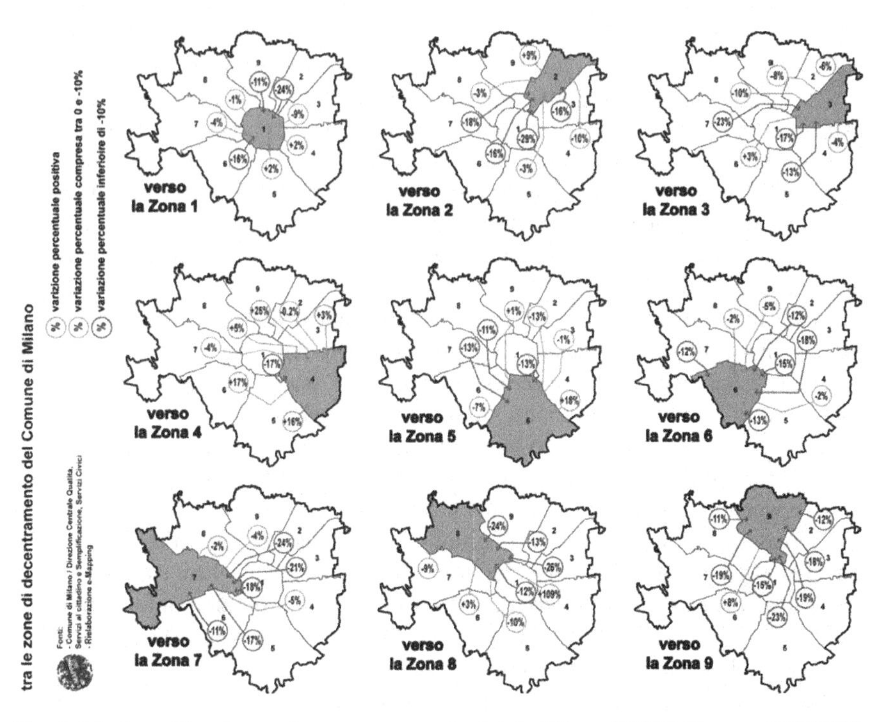

Fig. 6 Moving of immigrants from one area to another in Milan. ANCE Milano, e-MAPPING

The real challenge was to enter data with different contents: economic, demographic, social. Many maps where created as tools for developers: maps with the shifting of immigrants from one neighborhood to another and the growing of small commercial points shows the swing of prices in a slow movement across some years; other maps relate the age of population and the profession located in particular district around universities; another map shows public services such as schools o sports center as the catalysts of young families.

The tools of GIS software allowed us to use every ingredient as indicator to work on different urban responsive models.

The first trouble was the collection of dates useful to learn the essence of this territory. The culture of data sharing is at the dawn in Italy and very often privacy laws not really understood.

This was the occasion to have a good relation with public institution, to weave common programs with common goals.

During the first months people involved in the project spent time in collecting data sheet from everybody at least as a door to door work. Our aim was to have the largest quantity of dates giving us the opportunity to know our land. We spread agreements with Regione Lombardia, with public agency of Milano, with offices related to both health as culture.

The aim was the collection of every data to be able to describe urban transformation an social effect of it. At the beginning the purpose of our project seemed confused. The idea to be able to map contents with different origins and affiliations seemed incomprehensible to the most. The main difficulty was to explain the outcomes of our researches: a concise maps or sequences of maps; what years ago was called_urban analysis_now is presenting us as a computational science to explain urban phenomena. On a second step the necessity to ensure data privacy.

On a third the necessity to convince every actor to participate of a huge program of knowledge. Most of our interlocutors did not know the scope of parametric modeling of GIS, and used to elaborate summary tables that prepared for their summaries and statements without thinking to a possible georeferencing or homogeneity of the data.

Many boxes of data sheets were descriptions not fit for summary sheets complex. Many data were outdated and poorly configurable.

The research started with a series of agreements with agencies and institutions for access to archives of all kinds, from paper to digital.

Data that could give sense to territorial dynamics and instances settlement of a certain area were collected and cleaned from noises. Often the acquisition of an archive went on to failure owing to the fact that it was not born to be georeferenced and reported incomplete data, incongruous results of archiving. In the case of the office of Milan town hall, the natural place where all our data where supposed to be, the theme was hard to be organized. Offices dealing with building information and licenses do not have a single own data base that grows every day. Every technical manager collects data report on his own laptop.

There was not a typical shared report to guide the collections of information on building quantities. Final reports often aimed to explain charges for the client and not to give informations on buildings quantities. Some of the reports, as a paradox, had perfect economic counts but not a word on the building quantity.

The approval process of the practices also often lasts for several years; the last technical manager often changes data started by others at different times and there was no vision to change all the building data preparing file to be shared as guides by everybody. There is no culture on computational data as facts to plan.

We referred to data not yet normalized that we found in different offices in different forms: paper or word file, excel files, access, csv. Some data were for faster obsolescence but this is the reason why they often were recollected and evolved. This is the case of databases on real estate quantities connected with direct permits that we collected at the town offices. It is a changing data base. It summarizes, with daily scheduling, a situation that can be modulated according to transformative time thresholds useful in any match with other data providing a progress (Figs. 7, 8, 9).

The first aim of all this work was the transfer of knowledge to local developers to support their knowledge and allow economic policy of intervention.

In this regard we prepared different maps to show amount of volumes fielded by permission allowed by administrations.

These maps were accompanied by studies on the timing of realization, on the effective implementation of what was envisaged in relation to the time of estate

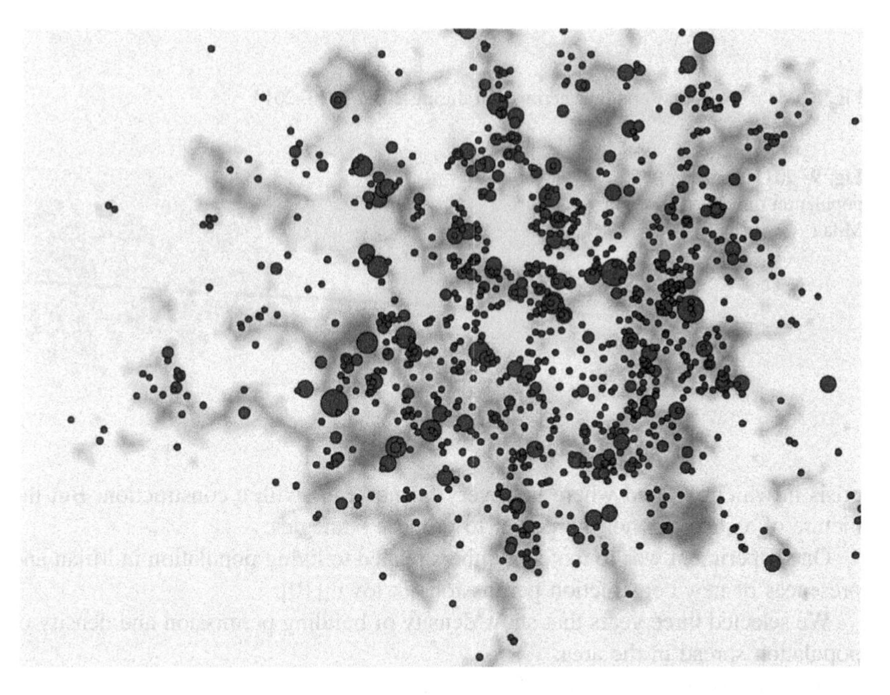

Fig. 7 Density of population in Milan (in the back) and presence of building permission during 2012–2013–2014. The circles show different dimensions related to the volumes supposed to be built

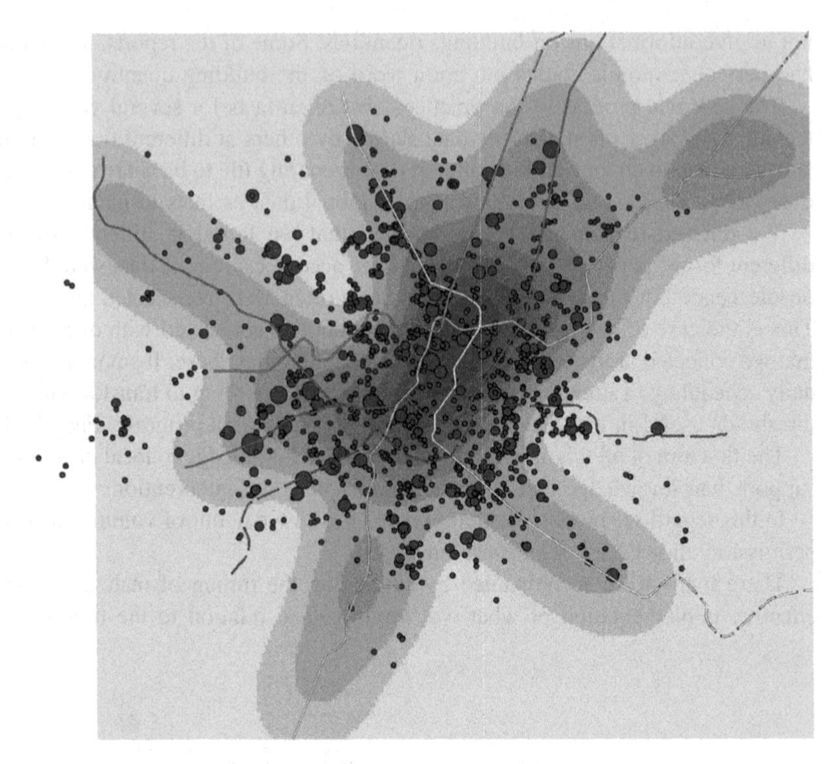

Fig. 8 Metro lines and building permission during 2012–2013–2014

Fig. 9 2011 Census of population on census grids in Milan

crisis in which we find where not every licence ends with a construction. But the picture of a state has not the power to propose strategies.

One experiment was to match numbers related to living population in Milan and presences of new construction permissions in town [10].

We selected three years that show density of building permission and density of population spread in the area.

Another experiment was the check of the connection between the location of the same building quantities and the transport such as metro in town. According to 2011 census we mapped density of population with small grids given by the census itself.

The main tool of this kind of research is often the census [11, 12], the last two are in 2001 and 2011. In addition to these, we collected data with other faster sources such as the opening of new businesses in the territory or the registration of new artisans or purchases of services to people that create situations not easily reading interpretable or even describable. In this regard we calculated intermediate situations, to mediate the possibility of computing different entities.

Mapping density of population with small grids given by the census itself did not prove to be the best parameter to evaluate any possibility of new real estate.

The aimed experiment for the developers was to see more interesting areas in terms of future development. We resorted to a spatialization of the data; also we introduced the real volumetric real estate data we had collected.

As a result we get a map, even if crude, showing the best opportunities for new development: places where the highest density of population compares with the lowest number of recent building (Fig. 10).

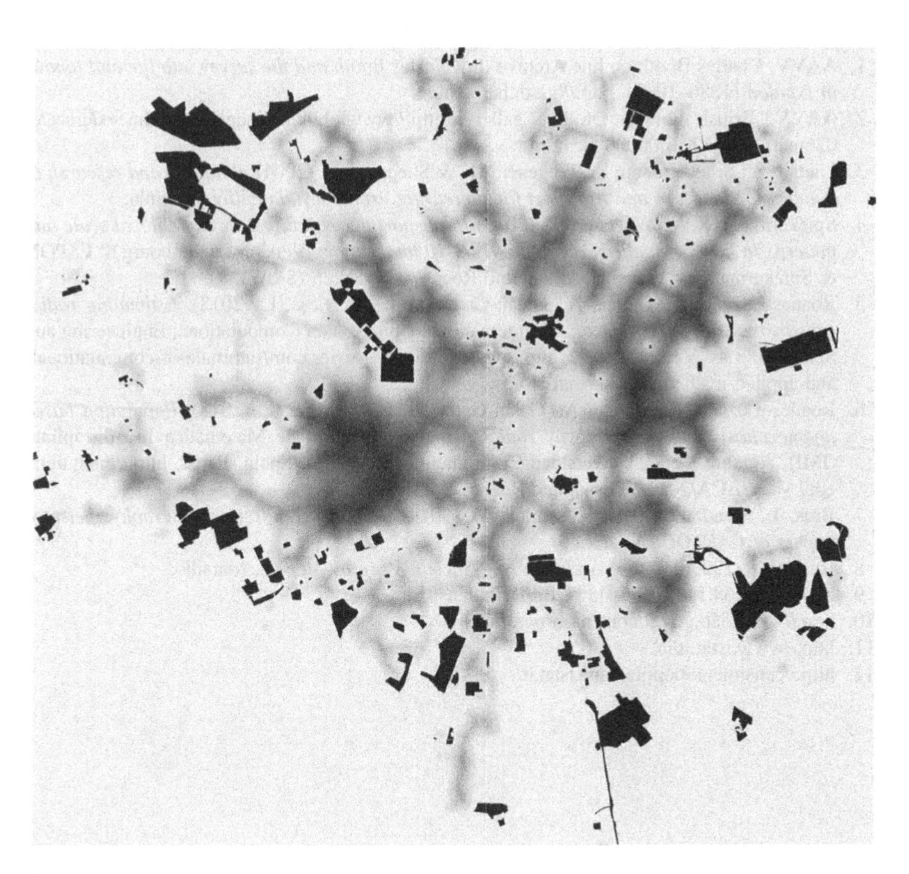

Fig. 10 Density of population and biggest real estate development

6 Conclusions

The representation of spatialized data is a certain base for environment of urban design. It currently assumes morphological characteristics of great interest, such as responsive maps, depending on the tools that are used for their cataloging georefering data belonging to different semantic categories is the right way to proceed to an analysis of the needs expressed by a city, whatever the needs of the population or a part of it.

In a world of open data more and more crowded of disposal ready to give us real time informations, GIS is an adequate computational methodology to study complex urban transformations and possibly suggest corrective how incentives. The matching of data set coming from different sources represent a real challenge in computational urban design.

References

1. AAVV. Charles Booth on line Archive. In *Charles Booth and the survey into life and labour in London* (1886–1903). http://booth.lse.ac.uk/.
2. AAVV. British library. On line gallery. http://www.bl.uk/onlinegallery/onlineex/crace/c/026map0000182c1u0000000c.html.
3. Lautso, K., Spiekermann, K., Wegener, M., & Sheppard, F. (2004). *Planning and research of policies for land use and transport for increasing urban sustainability*. Helsinki.
4. Spiekermann, K., & Wegener, M. (2015). *Transport accessibility at regional/local scale and patterns in Europe applied research 2013/1/10 final report*. Version Luxembourg © ESPON & Spiekermann & Wegener, Urban and Regional Research (S&W).
5. Roanes-Lozano, E., Garcia, A., Galán-Garcia, J. L., & Mesa, L. (2013). *Estimating radial railway network improvement*. 4th International Congress on Computational Engineering and Sciences, Las Vegas, Paper. http://www.journals.elsevier.com/saturnale-of-computational-and-applied-mathematics/special-issues.
6. Roanes–Lozano, E., Garcia, A., Galán-Garcia, J. L., & Mesa, L. (2013). *Population based anamorphosis maps for railway radial networks*. Instituto de Matematica Interdisciplinar (IMI), Algebra Dept., Universidad Complutense de Madrid, Spain, Paper. http://math.unm.edu/~aca/ACA/2013/Nonstandard/Roanes-Lozano2.pdf.
7. Brus, J., Vondrakova, A., Vozenilek, V. (2014). *Modern trends in cartography: Selected papers of CARTOCON*.
8. Burdett, R. (2006). *Città: Architettura e Società Venezia*. Venice: Marsilio.
9. http://109.168.114.139/assimpredil/Default.aspx.
10. http://www.dati.gov.it/content/infografica.
11. http://www.istat.it/it/.
12. http://censimentopopolazione.istat.it/.

Part III
Digital Applications and Cultural Heritage

3D Modeling of an Archeological Area: The Imperial Fora in Rome

Tommaso Empler

Abstract In disseminating and promoting the understanding of the cultural heritage, 3D modeling is increasingly used to provide an idea of what objects actually looked like in the past. The procedures and tools used for this purpose are familiar from their applications in other areas. This paper presents a method for creating a 3D modeling of an archeological Area: the Imperial Fora in Rome using an open-source pipeline. In addition to providing tools for generating and controlling forms, the modeler also makes it possible to navigate through the textured model in real time. The model can be 3D printed to produce a physical version of the architecture of the past.

Keywords 3D modeling · Cultural heritage · Real time · 3D printing

1 Introduction

The various ways of disseminating and achieving an understanding of the cultural heritage, depending on the tools used for visualization by scholars or those who are simply curious about the legacy of the past, often need appropriate 3D reconstruction models, where the type of information, and consequently the modeling method, may vary as regards the level of detail and the number of polygons used in the 3D model, which must be appropriate for the devices the models will run on.

The Imperial Fora in Rome are among the most extensively investigated archeological sites, where the studies conducted by the central heritage institutions (the Soprintendenza di Stato and Sovrintendenza Capitolina), research organizations (universities in Italy and abroad and recognized research centers) and scholars have generated an impressive legacy of 3D documentation.

T. Empler (✉)
Department of History, Representation and Restoration in Architecture,
Sapienza University of Rome, Rome, Italy
e-mail: tommaso.empler@uniroma1.it

M. Rossi and G. Buratti (eds.), *Computational Morphologies*,
https://doi.org/10.1007/978-3-319-60919-5_14

The paper discusses the various possibilities for 3D modeling and visualization, and suggests further methods that employ procedures derived from other disciplines or areas of application, but which are as yet little used for the cultural heritage. The Forum of Nerva is taken as a case study.

2 Recent Solutions

Of the many 3D reconstructions of the Imperial Fora, we have considered "*Rome Reborn*" and "*Progetto Traiano*": the first as an example of a joint study by universities and research organizations, the second as a study carried out by private specialists in three-dimensional modeling.

Rome Reborn

Rome Reborn[1] is an international initiative whose goal is to create 3D models illustrating ancient Rome's urban development from the first settlements in the late Bronze Age (ca. 1000 B.C.) to the depopulation of the city in the early Middle Ages (ca. 550 A.D.).

The initiative got under way in 1997, when the Virtual World Heritage Laboratory of the University of Virginia (VWHL) joined forces with the UCLA Experiential Technology Center (ETC), the Reverse Engineering Lab at the Politecnico di Milano, the Ausonius Institute del CNRS, the University of Bordeaux-3 and the University of Caen in collaborating on the creation of a digital model of ancient Rome as it appeared in late antiquity. The notional date of the model is June 21, 320 A.D. Since 2009, the sponsor and administrative home of the project has been Frischer Consulting, whose mission is to apply 3D technologies to the study and dissemination of cultural heritage throughout the world [1, 4].

The primary purpose of the project is to spatialize and present information and theories about how the city looked at that moment in time, which was more or less the height of its development as the capital of the Roman Empire. A secondary goal was to create the cyberstructure whereby the model could be updated, corrected and augmented.

The digital model reflects the sources of our knowledge about ancient Rome, which are of two kinds: archeological data about specific sites and features ("Class I"); and quantitative data about the distribution of building types throughout the fourteen regions (or wards) of the city ("Class II"). Features in Class I are known from archeological excavations and studies; coins; inscriptions; ancient literary sources; and artists' views from the Renaissance until the nineteenth century. Features in Class II are known from two regionary catalogs (the Curiosum and the Notitia) dating to the fourth century A.D.

The digital model as a whole consists of two types of materials: highly detailed models of buildings that can be reconstructed on the basis of reliable archeological

[1] http://romereborn.frischerconsulting.com/about.php.

evidence (examples: the buildings in the Roman Forum and the Forum of Julius Caesar; the Flavian Amphitheater; the Temple of Venus and Rome, etc.); buildings and other features that are known only by type and by frequency in the particular regions of the city. Approximately 200 buildings of the first type and between 7000 and 10,000 in the second category have been modeled. Around 50 of the 200 Class I buildings that were modeled have been created with the help of scientific advisory committees of experts, while detailing operations are still to be completed for the remaining 150. The Class II buildings have been modeled by a procedure entailing the digitization of the Plastico di Roma Antica created by Italo Gismondi from 1934 to 1974.

As regards dissemination, videos of the digital model have been posted for viewing starting in June 2007. In 2008, the 1.0 version of Rome Reborn was published on the Internet as "*Ancient Rome 3D*" in Google Earth. In 2012, this layer was removed. From 2012 to 2013, a number of initiatives surfaced to use Rome Reborn 2.2 as the major asset for educational videos, whether shown in museums or on the Internet. Meanwhile, research was pursued by the leadership of the project on ways of making the model interactively available through a game engine.

Progetto Traiano

Progetto Traiano[2] is a collaborative project involving both engineers and archeologists which uses modern three-dimensional design technology to give form to the findings and observations stemming from studies of the monuments of ancient Rome. Attention has focused on creating reconstructions that are as faithful as possible in terms of construction volumes, size of elevations, and structural feasibility. At the present state of the project, however, the reconstructions cannot claim to be reliable as regards appearance, and the colors and surface finishes shown are thus purely evocative in intent.

The reconstructions are based on a thorough critical analysis of documentary material from the various Sovrintendenze, including photographs and publications, all of which were compared with field surveys. As regards the volumetric aspects, the models in Rome's Museo della Civiltà Romana were used as a starting point, and were analyzed and reinterpreted in the light of recent studies.

All reconstructions, animations and film clips were creating using only open-source software (Linux, Blender, YafaRay, Gimp).

3 The Research

3D Modeling Tools

Three-dimensional modeling and visualization of the Imperial Fora involved an initial stage in which the information for reconstructing the Fora was acquired from survey and excavation campaigns, studies by researchers and historians,

[2]http://www.progettotraiano.com/.

iconographic sources and archival documentation [3, 5–8]. This was followed by a second stage in which the documents were digitized and vectorized so that they could be used in a digital environment: hardcopy documents were scanned, manipulated with a raster image editor, and then converted to 2D vector using a CAD application. In the third stage, a three-dimensional modeling program was used to transform 2D information into 3D, while the final stage consisted of producing static and dynamic images that give a reliable view of the Fora's appearance in the Imperial Roman period.
The entire pipeline used freeware programs:

- Photo retouching and editing of the raster files supplied by the Soprintendenza and post-production of the files generated with the rendering procedure: GIMP (open source).
- CAD for vectorizing and 2D survey file editing: DraftSight (freeware).
- 3D modeling and frame rendering for final visuals: Blender (open source).

Image Editing—GIMP
GIMP, which stands for GNU Image Manipulation Program, is a cross-platform photo editing program.

It is a valid alternative to applications like Photoshop, and users who are familiar with the latter will have no problem passing to GIMP, which features the same philosophy and palette of controls, as well as very similar filters. It is a highly flexible application that can be used as a simple paint program, an expert quality photo retouching program, an online batch processing system, a mass production image renderer and an image format converter.

CAD—DraftSight
DraftSight is professional 2D CAD software produced by Dassault Systemes for users who want to create, edit and view DWG, DXT and DWT files. The program contains all the standard management functions used in the sector (Cartesian coordinate system, layers and layer manager, background masks for note, etc.), drawing functions (through the most common primitives), and transformation functions. As freeware, DraftSight requires activation.

Modeler/Renderer—Blender
Blender[3] is a 3D graphics and animation software. It can be used for modeling, rendering, nonlinear editing and video compositing, and can implement modules for creating interactive 3D applications.

Blender is an open-source program with a wealth of the features typical of advanced modeling systems.

These features include:

1. Supporting an enormous variety of geometric primitives, e.g., polygon meshes, Bézier curves, NURBS surfaces, metaballs, digital sculpting and vector fonts.

[3]Empler [2], pp. 68–70.

2. Tools for managing key-framing animation, such as the use of forward/inverse kinematics, soft and rigid body dynamics, fluid simulation and collision detection, and a particle system for simulating hair and collisions between objects.
3. Basic non-linear video editing.
4. The Blender Game Engine, which manages collision detection, dynamics simulation and logic programming, making it possible to create standalone programs or real-time applications that range from viewing architectural elements to creating videogames.
5. Two internal rendering engines, the first being a fast, versatile engine whose many features include radiosity algorithms, render baking to UV maps, edge rendering for toon shading (for pencil sketch rendering) and ambient occlusion. The second is the Cycles Render Engine released with version 2.60, which uses more advanced algorithms to achieve true photorealism thanks to an accurate reproduction of the physical laws that govern the actual behavior of light and, unlike most rendering engines, uses the GPU instead of the CPU to compute the end product. Consequently, it takes only a few dozen minutes to produce photorealistic visuals that only a couple of years ago would have required hours, if not days. It is also possible to integrate biased and unbiased external engines such as YafaRay and LuxRender, which are also open-source, as well as proprietary engines like Indigo and Octane.
6. Python scripting to automate or control many aspects of the program.

Three-Dimensional Reconstruction Methodology

The reconstruction of the Forum of Nerva in the Imperial Roman period is based on the graphical layout proposed by the archeologist Roberto Meneghini in his book "*I fori imperiali. Gli scavi del comune di Roma 1991–2007*", which hypothesizes a possible configuration of the plaza without advancing theories about the parts that are still uncertain, such as the western termination of the Forum [9].

For the elevated parts, the survey of the Colonnacce drawn up by the Sovrintendenza was used, and was entered in scale in the CAD application for subsequent vectorizing.

After entering the Forum's ground plan and section in the three-dimensional modeling program, the structural components of the model were constructed.

The first operation was the extrusion of the perimeter walls that constitute the Forum's physical boundaries, using the heights of the various elements in the section.

The second step was to extrude the columns, which called for more detailed modeling because of the cornices making up the entablature, which projects from the wall behind the shafts, and the capitals (Figs. 1, 2).

The colonnade of the monumental plaza was put together from the detailed model of a single column, its capital, the cornices and the projecting entablature (Fig. 3).

The single column used as a starting point was duplicated using the modeler's array function, which makes it possible to reproduce an object at precise distances

Fig. 1 Mesh 3D of the trabeation of the Forum of Nerva (Image by Burda Klit)

Fig. 2 Texturing of the trabeation of the Forum of Nerva (Image by Burda Klit)

along three Cartesian axes. The replicated objects are identical copies of the original and are linked together so that modifying any one of them will modify them all at the same time. The entablature between each column and the next was produced by creating a connecting band which was then replicated between all of the shafts.

In reconstructing the Temple of Minerva, it is necessary to have a knowledge of certain generic features that commonly recur in Roman temples, such as the type of monumental door, the pavement, the roofing and the probable, though as yet unascertained, presence of acroteria.

The Temple was modeled on the basis of the historical iconographic sources that have come down to us today, while the portions that have survived were completed using assumptions based on probable similarities and correspondences with the site. This is the case, for example, of the pediment of the Temple of Minerva (Fig. 4), the few remaining fragments of which do not permit a definitive reconstruction of

Fig. 3 3D Modeling of the Perimetral wall of the Forum of Nerva (Image by Barbara Forte and Emanuele Fortunati)

Fig. 4 Reconstruction view of the Forum of Nerva from the Temple of Minerva (Image by Barbara Forte and Emanuele Fortunati)

the original architecture. The major question in reconstructing the pediment regarded the frieze with the dedicatory inscription and its relationship with the moldings on the side entablature. After first hypothesizing that the trabeation extended across the entire front of the temple, connecting the two side branches as suggested in a historical document that has come to light, we opted for a "*modus operandi*" found in other structures of the same period such as the Temple of

Vespasian, where the frieze takes up the entire height, including the trabeated portion, and includes the inscription under a cornice with palmettes. Once this point had been decided, the problem arose of how to resolve the two frontal corners of the side entablature, consisting of several stepped moldings with the smooth portion bearing the inscription. This issue was clarified by considering the historical representations by Dosio and Palladio, which illustrate how the two elements were related, with the trabeation extending to the front of the temple, but only for a small space, thus connecting the sides without interfering with the inscription.

Visualization

Once modeling was completed, visualization was the most innovative and interesting part of the project.

The procedures used are well-known, but have not yet been extensively applied to convey information about archeological areas: real-time navigation and a physical model of the reconstructed Forum produced with a 3D printer.

Real-Time Navigation

As Blender's programming structure includes a game engine capable of running simultaneous multiple events, three-dimensional modeling with this program makes it possible to take advantage of a number of features that permit real-time visualization, including:

- Scene rendering with texturing and light effects
- Physics engine
- Management of sound events
- Management of source code scripts
- Animations.

Together, these features permit "subjective" movement simulation, which is achieved with single images produced in quick succession, e.g., at 30–60 fps (Frames Per Second), thus providing a smooth visualization.

A navigable model must achieve an effective tradeoff between the number of polygons in the 3D model and the processor's ability to run a sufficient number of frames per second ensure that movement is always smooth. The model must thus be optimized for the viewpoint of the "*virtual camera*" located at around 1.70 m above ground level. Thus, the objects nearest the camera "*viewpoint*" are modeled in greater detail, while those that are farther away are carefully textured using shadow maps, bump mapping, specularity and diffuse mapping. Layering different textures provides a very high level of photorealism, and at the same time simplifies computing operations.

The shadows in the model are computed by setting the light sources. As continuously computing the illuminated and shadowed areas in real-time would slow down the navigation mesh, a "*baking*" process is used to pre-compute the shadows that the light source casts on each physical object in the model. This tool renders only the shadows on a specific surface, and enables them to be included among the textures to be applied to an object.

After the 3D model for the Imperial Roman period was created, the Blender Logic Editor was used to set up and edit the game logic for the objects present on the scene in real-time. The logic for the objects created and selected in the 3D panel consists of logic bricks, which are shown as a table with three columns, indicating sensors, controllers and actuators. The links joining the logic bricks conduct the pulses between sensor-controller and controller-actuator and permit the physical actions that take place in the scene, indicating the direction of the logical flow between objects.

The game properties are the variables used to store and access the data associated with each modeled object.

The model of the reconstructed Forum of Nerva is exported in a standalone.exe player file which allows the model to be run without having to load the program that generated them, in this case Blende.

The structure of the navigation interface was also developed with the Blender Logic Editor. Its objectives are to guarantee smooth navigation mesh movement through the keypad and joypad (like the Dualshock4 wireless controller for PS4), make it possible to pass quickly between two models if necessary (i.e., one for Imperial Rome and one for a later period), and provide interactive information hotspots that can be accessed for all parts of interest in the 3D model. Hardware controls can be interfaced with input commands with the aid of several logic diagrams and scripts.

The system whereby the end user interfaces with the application includes a main menu, where the user can chose to begin navigating in person in one of the two historical periods, two models in axonometric view with the information hotspots that can be found while navigating through the model, and a screen with a map of the commands that can be used.

3D Printing

The other visualization method consists of a 3D printed model reconstructing the Colonnacce, the one section of the Forum of Nerva's perimeter wall that is still standing.

Three-dimensional printing uses the CAD/CAM procedure (where the 3D model is exported as an.stl file), with a virtual division of the object into sections or parts to be sent to the printer, as required by the type of printing (employing polylactic acid, or PLA), which does not permit voids to be left (Figs. 5, 6).

Fig. 5 3D printing reconstruction: the "Colonnacce" (model by Mattia Fabrizi, 3D Printing By CeSMA Laboratory)

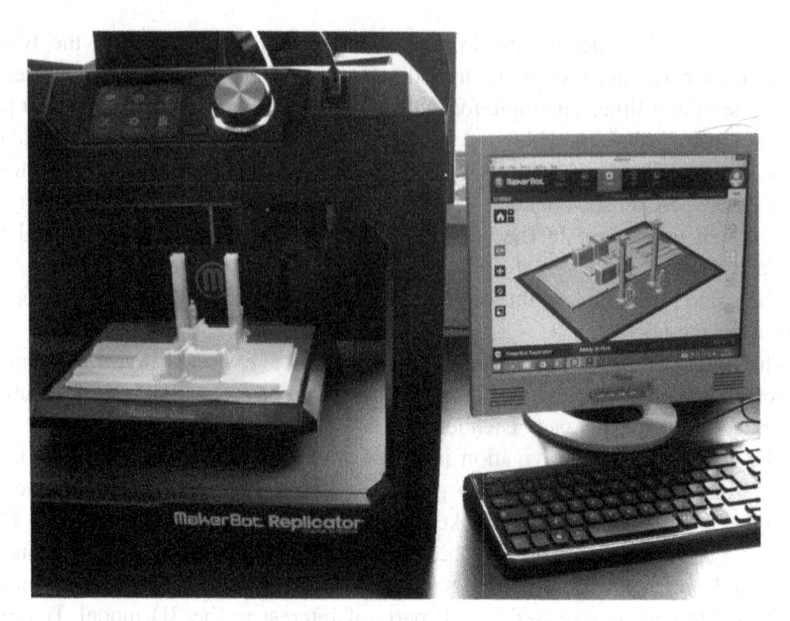

Fig. 6 3D printing process of the "Colonnacce" (model by Mattia Fabrizi, 3D Printing By CeSMA Laboratory)

4 Conclusion

Being able to use visualization tools with real-time navigable 3D reconstructions and/or 3D printed physical models opens up new and interesting prospects for communicating the cultural heritage: scholars can verify hypotheses regarding reconstruction and/or modify them in real-time in the digital model, while visitors can gain an understanding of what a site actually looked like in the period of its greatest splendor.

In the future, the 3D reconstruction of the Forum of Nerva will be extended to the entire area of the Imperial Fora, both in real-time navigable form and as a 3D printed "*physical*" reconstruction. For real-time navigation in particular, the model will also have to be usable with smartphones and tablets.

References

1. Dylla, K., Frischer, B., Mueller, P., Ulmer, A., & Haegler, S. (2010). Rome Reborn 2.0: A case study of virtual city reconstruction using procedural modeling techniques. In B. Frischer, J. Webb & D. Crawford (Eds.), Making history interactive. Computer applications and quantitative methods in archaeology (CAA), *Proceedings of the 37th International Conference, Archaeopress, Virginia.*
2. Empler, T. (2008). *Software libero per la progettazione*. Roma: Dei.

3. Fanini, B., Demetrescu, E., Ferdani, D., & Pescarin, S. (2012). Aquae Patavinae VR, dall'acquisizione 3D al progetto di realtà virtuale: una proposta per il museo del termalismo. In *Atti del Convegno internazionale "Aquae Salutifere", Università di Padova e Soprintendenza dei Beni Archeologici del Veneto, Padova*
4. Guidi, G., & Frischer, B. (2005). Virtualizing ancient rome: 3D acquisition and modeling of a large plaster-of-paris model of imperial rome. In A. Beraldin, S. El-Hakim, A. Gruen, & J. Walton (Eds.), *Videometrics VIII*. San Jose: SPIE.
5. Meneghini, R. (1991). *Il Foro di Nerva*. Rome: Fratelli Palombi.
6. Santangeli Valenzani, R. (2001). Il Foro di Nerva. In M. S. Arena et al. (Eds.), *Roma dall'antichità al medioevo: archeologia e storia nel Museo Nazionale Romano Crypta Balbi*. Milano: Electa.
7. Santangeli Valenzani, R. (2000). I Fori Imperiali in età post-classica: i Fori di Cesare, Nerva e Pace. In S. Baiani & M. Ghilardi (Eds.), *Crypta Balbi – Fori Imperiali, Archeologia urbana a Roma e interventi di restauro nell'anno del Grande Giubileo*. Roma: Kappa.
8. Ungaro, L., Meneghini, R., & Milella, M. (2014). *Le Chiavi di Roma. La Città di Augusto*. Roma: Edizioni CNR.
9. Viscogliosi, A. (2000). *I Fori Imperiali nei disegni d'architettura del primo Cinquecento: Ricerche sull'architettura e l'urbanistica di Roma*. Roma: Gangemi.

Augmented Visualization: New Technologies for Communicating Architecture

Alberto Sdegno

Abstract The aim of this research was to define a procedure to communicate complex digital model of architecture using Augmented Reality (AR) algorithm. This technology, in fact, could be very useful to transmit results of experiments that are dense of references to be immediately understood, due to the interaction of the user involved directly in the process of understanding. The application was developed both for simple architectural models—such as small rooms—and to complex morphologies—which have, for example, detailed capitals—to test the way in which the user is involved into the system.

Keywords Representation · Augmented reality · Visualization · Architecture · Digital model

1 Introduction

The Augmented Reality is based on the concept that every object can be associated to a set of information which can describe it in a multiplicity of ways of knowledge. So, the object itself refers to a lot of levels of communication to help the user to have structured notes on it, in reference to some aspects that were to be transmitted. The AR procedures can be applied only to digital systems, that are the way in which information can be superimposed without limitation, and can be distributed worldwide thanks to the potentiality of the web. We can use indifferently images, videos, models, textual notes, that can be linked each other to augment the knowledge of the user, who can interact with them with his/her own hands, standing in front of the cam (Fig. 1).

We can find the origins of AR in many essays and experiments that have something to do with the association of different levels of information. One of the first is without any doubt the Memex, that was a system invented by Bush [1] in the

A. Sdegno (✉)
Department of Engineering and Architecture, University of Trieste, Trieste, Italy
e-mail: sdegno@units.it

M. Rossi and G. Buratti (eds.), *Computational Morphologies*,
https://doi.org/10.1007/978-3-319-60919-5_15

Fig. 1 Application of AR to a digital model of the picture *The feast in the house of Levi*, Paolo Veronese, 1573

30's of XX Century, but published in 1945, in which the researcher described a machine which was able to archive and find some different kinds of data, both alphanumerical and graphical. As at that time there were no digital system already developed, it remained as a theoretical prototype similar to the hypertext, which, as we know, was created some years after. Another interesting machine was the Damocles' Sword developed by Ivan Sutherland in 1965—and published in 1968 [2]—which anticipated the Virtual Reality Systems and, in a certain way, AR ones. In fact it was based on a head mounted display which allowed the user to see a wireframe model, having the possibility to turn around it at every movement of the head. So the user could interact with it, making him more friendly to the object he is perceiving, although it is not a real one.

At the same conference in which Sutherland presented this system—the 1968 Fall Joint Computer Conference—Douglas Engelbart spoke about a way to "*augment human intellect*" [3]. The term "*augment*", associated to new technology, was already used, but in this case it is very similar to the meaning used for AR. In fact Engelbart in this occasion showed for the first time the use of the "*mouse*", an instrument to simplify the interaction with data. In detail the author told about the development of "principles and techniques for designing an "*augmentation system*".

But it was around 1990 that the technology allowed to experiment some algorithm similar to the ones that now characterize the AR systems. Probably the term 'Augmented Reality' was used for the first time by Tom Caudell working on a Boeing's Computer Service' Adaptive Neural Systems Research and Development project in Seattle, trying to find an easier way to help the aviation's company's manufacturing and engineering process, instead of using diagrams and marking devices.

Some other important researches led to the present development, which is now structured in very different ways to interact with reality using a digital display.

2 Augmented Reality for Architecture

The research we are developing in the field of AR is strictly dedicated to architecture. The answer was if we can use AR also thinking to the characterization of the world of architectural models, which, as we know, have a great differentiation inside their typology. In fact it is very simple to interact with small rooms, to show them as if they were only boxes, without the specification of lighting, materials, furniture, and so on. The file to be processed is simple and you can treat it without great difficulties. But if you decide to give realism to the virtual object you could have in your hand, the process can be much complicated, because every element have to be treated separately, in order to satisfy the conversion procedure.

There are more possibilities of investigating digital models with AR.

The simpler was associating a QR-code (QR-code is for Quick Response Code) with a model and then to view it when you show the QR-code to the web cam of a computer. The transformation is immediate: moving the printed code in your hands the digital model you see on the monitor will move in real-time. Depending on the algorithm you use, the model can be manipulated during the operation of transformation. For example it is possible to slice a model, to visit (from the exterior to the interior), to change the properties of color of the wall, etc. (Fig. 2).

As the QR-code is a quick way to identify some data without ambiguity—similarly to a barcode, but with an extended set of information—it was developed a new way to associate data, without using an abstract code. The name is Markerless Augmented Reality, that is an AR technology that can extract from an image (but potentially also from a part of the environment) the information useful to be

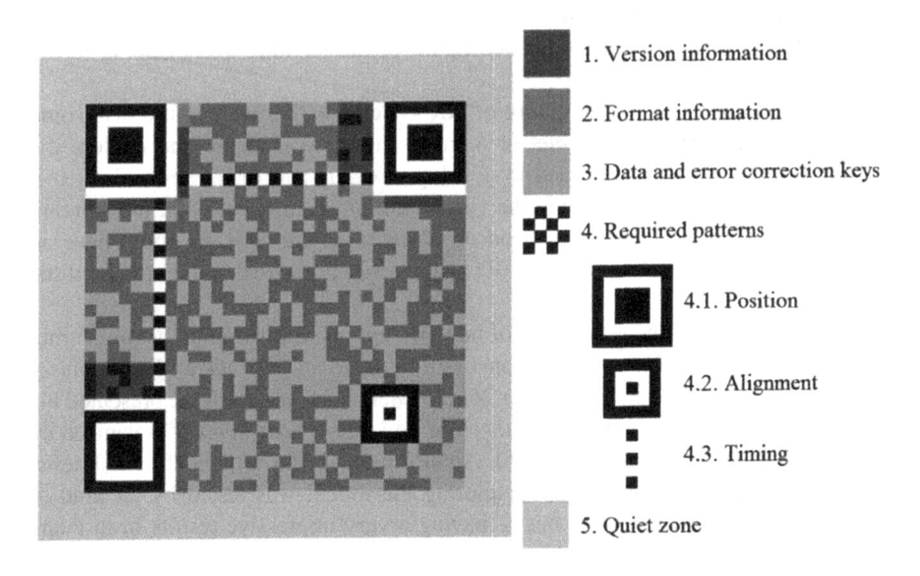

Fig. 2 The structure of a QR-code

elaborated into an AR system. Some algorithms, in fact, are able to identify an image, considering it as if it was a sort of code, and then associate it to a digital model. Without any doubt this is a more direct way to enable the user to understand what kind of data he is going to treat for AR. Instead of QR—codes—that although are all different, they are very similar and seem to be the same—the code based on an image offer a more specific approach to the operation. Being studied for the development of interactive videogames, the Markerless AR can be more friendly and the user can be involved to practice it as if he were playing a game.

3 A Painting by Paolo Veronese in AR

After the theoretical investigation we decided to apply the AR algorithm to an architectural example, to be used above all during exhibitions and public sessions. So we try to prepare a 3D file of a complex architectural scene, extracted from a sophisticated perspective analysis of a well known painting from the history of art—the work is "*The feast in the house of Levi*", painted by Paolo Veronese in 1573 [4]—in order to simulate the entrance into the scene by the observer standing in front of it.

As the use of Augmented Reality needs an optimization of the geometry of the 3D model, we needed to interpolate the most complex form—such as the capitals—with the LOD (Level of Details) algorithm, to have a final 3D interactive model which could be easily explored in space.

After this there was the phase of mapping all the materials of the objects with appropriate textures, and then, to try to represent the scene in a realistic way. All the people, in fact, were modeled in a silhouette 2D vertical form, having care of extracting the texture by the reference figure of the painting and mapping to the surface.

It is true that the real potentiality of AR systems is that we can use also some portable devices, such as a tablet or a smart-phone, to navigate and explore the interactive model. So we decided to develop also this part of the research as if it were in front of a work of art and the visitor of a museum could point the painting on the wall with a tablet and, instead of taking a picture of it, he could have an added value from it, transforming in 3D and perceiving the real configuration of the space.

We tried to experiment this kind of new perception with common people during an event that took place in occasion of "*Trieste Next—European Exhibition of Scientific Research*", from September 26th to 28th, 2014, in which we presented the system inside the space called "*Architecture and Augmented Reality*" organized by the Department of Engineering and Architecture of the University of Trieste. The simplicity of the method for exploring the model was tested by all kind of persons—from children to old ones—having a very impressive results from them.

Fig. 3 Digital reconstruction of the scene with figures

Fig. 4 Public presentation of the AR prototype

Finally we have to remember also that last upgrades of the software allow to consider the movement of some object inside the scene, so that the static scene represented in the painting could be transformed into a movie or theater set (Figs. 3, 4).

Acknowledgements The main research on AR was developed at the Department of Engineering and Architecture and was financed by the University of Trieste (F.R.A. 2013, Fondo di Ricerca di Ateneo) titled "*Augmented Architecture*", directed by Alberto Sdegno. The research on Veronese's painting was financed by the 2010-11 PRIN Research from the Italian Ministry of Education, University and Research titled "*Architectural perspective: digital preservation, content access and analytics*" (Prot. 2010BMCKBS), directed by Riccardo Migliari (University of Rome "La Sapienza"). The model of the scene was made by PhD student Silvia Masserano, the AR application was made by architects Denis Mior and Eleonora Gobbo.

References

1. Bush, V. (1945). As we may think. In *Atlantic Mounthly*, 176, July 1945 (pp. 101–108).
2. Sutherland, I. E. (1968), A head-mounted three dimensional display. In *Fall joint computer conference, AFIPS onference proceedings*, Vol. 33, Part 1, San Francisco 1968 (pp. 757–764)
3. Engelbart, D. C., & English, W. K. (1968). *A research center for augmented human intellect*. In Fall joint computer conference, AFIPS conference proceedings, Vol. 33, Part 1, San Francisco (pp. 395–410).
4. Sdegno, A., Masserano, S. (2014). *The inverse perspective for analyzing a painted architecture* by Paolo Veronese. In *Geometrias'14, proceedings of XIII aproged conference*, Lisboa (pp. 31–37).

The Video Animation: An Innovative Way to Communicate

Nicola Velluzzi

Abstract Communication is increasingly becoming one of the primary goals of architecture and design. The need of searching effective strategies for interaction with society and that of sharing knowledge and experiences are growing day by day. This study aims the creation of a new communication channel which can be able to combine contents with the above mentioned outreach strategies. Videos make unexperienced users to approach to a scientific products which remained exclusive tool for advanced experienced persons for long time. So, videos have been able to change the way of approaching to the scientific method.

Keywords Communication · Video · Animation · Masaccio · Trinity

1 Introduction

The prevention of users from cryptic languages and poorly understandable drawings is very important today in the architecture field. Several studies have shown that videos are becoming the predominant communication medium especially on Internet platforms. Such phenomenon is due to the peculiarities of the video, which is characterized by the simultaneous use of multiple semantic resources, such as images, languages and sounds. In this way, it creates visual propositions which added to the verbal ones. Therefore, the language becomes the focal point of communication. In this regard, it is appropriate to refer to the linguist Jakobson's[1] statements on the correspondence of the communication factors and the functions of language. So the conative function[2] assumes a relevant importance, to which

[1]R. Jakobson, Saggi di linguistica generale, Feltrinelli, Milano.
[2]The conative function (from the Latin conari = groped, search) or persuasivefunction is one that tending to persuade, to inffiuence, to convince the recipient.

N. Velluzzi (✉)
Department of Architecture, Università di Firenze, Florence, Italy
e-mail: nicola.velluzzi@unifi.it

M. Rossi and G. Buratti (eds.), *Computational Morphologies*,
https://doi.org/10.1007/978-3-319-60919-5_16

Jakobson link the persuasion of the recipient [3]. The persuation of the recipient and the spreading of a new scientific data through innovative ways, different than the traditional written text.

The idea of using video as a tool of communication has to rely to an innovative element: computer animation. It has the purpose to obtain theoretical and cultural reflections recognizing the real potential of content expression. In this way users can transfer through a dynamic image what until now had been transferred through the image of drawing. It should be noted that the new tool does not replace the other one: the objective is only to facilitate the acquisition of concepts and allow viewers to enter into an understandable and actual "space".

2 The Research

The purpose of this research, although with the knowledge to deal with a topic in continuous development, is to explain, as fully as possible a case study: the use of animation in the understanding of the perspective drawing in Masaccio's Trinity.

The current knowledge on perspective comes from what the didactic has handed down over the years. It is possible to think that these notions do not reflect at all the original concept. However is it possible to assume that what it was originally designed has a different nature? The animation suggests an answer to this question, opening a new "perspective" on the problem. Videos use the three-dimensional geometric occurrences in a more realistic way by separating from the drawing board. This process is probably connected to the existential experience of the artist. Many studies have been realized for interpreting the perspective drawing of Masaccio's Trinity.

They assumed as certain that the painting had inside some elements for adjusting itself in order to improve the entire result of the work. It was considered "practically" impossible that the painting had no projective lacks, even when the rule was applied rigorously. Therefore it was considered an approximation adopted by the painter. It's important to consider that the analyzed studies (especially those realized before the restoration of the painting) were based on reproductions different from the original with generally different proportions. As result it has been not easy to get plausible deductions. Thanks to the evolution of digital photography and the sampling process of images, the subsequent studies have been able to use a superior quality material by reaching more reliable results.

The research realized in a recent study, has focused in particular on the reading of three prominent studies aimed at the reconstruction of the perspective scheme of the Trinity by Masaccio. Chronologically analyzes the interpretations of Marisa

[3]Jakobson schematizes six key aspects identify a sender, a message, a recipient, acontext, a code and a contact and connecting them respectively the emotionalfunction, the poetic function, the conative function, the referential function, themetalinguistic function and function effortlessly.

Dalai Emiliani, Martin Kemp and finally the latest Filippo Camerota. In his study on the Trinity, Marisa Dalai Emiliani analyzed some previous researches at identifying the observer's distance point. In particular cites the Kern's thesis (1913)[4] and Degl'Innocenti's thesis: the first put the observer at a distance of 6.12 m; the second say that there is a band space which extends along the horizon line up to 3.85 m from the surface of the wall, the point from which it is possible to obtain an overall perception of the picture. Degl'Innocenti calculates the difference between the height of the horizon line (1.71 m) and the lower end of the columns: the result, just over a meter (1.06 m), corresponds to the height of a tall man about 1.70 m, kneeling on a platform as high as the base of the columns painted. He deduce that, if into this space was inserted an authentic altar at the same height as that probably painted, the picture would be perfectly perceivable from a perfect celebrant kneeling. Kemp focus its study on the top of the picture, in particular above the plane where the work's clients are kneeling. Immediately he notices that all the orthogonals of the vault converge on an only convergence point (V), identified just below the floor where the two figures kneeling. Kemp make the perspective theme inversely, and that need the identification of a figure with geometric characteristics known and the study of its deformation. So try a primary figure, a square or a cube, he reclaim the Kern's hypothesis on the basis of the covered area of the barrel vault is that of a perfect square. Kemp draws, therefore, a square in perspective with the sides that join to the convergence point (V). The upper side of the square is put on the start of the arc and he specifies that to identify, on the central perpendicular, the precise points where they are the centers of the arches of the barrel vault there are some ways. The mode consider more reliable is the one that makes use of an auxiliary construction in which the calculations are derived "from side" and not operating on a longitudinal section (Figs. 1, 2, 3).

The Kemp's construction leads him to assume even the exact point from which you must observe the painting. It is derived from the diagonal of the square, before quoted, extended until it meets the prolongation of the horizon line passing through the convergence point (V). The distance from the intersection point and V give the distance of the point of view from the picture plane. This result is a good approximation for Kemp equivalent to twice the width of the picture, and very similar to the width of the aisle. Then Kemp produces a list of irregularities that he sees in the picture: the lacunar's curved outline are not symmetrical compared to the design of the arches engraved that are supposed to indicate the midline; the width of the ceiling coffer is larger than the other six that make up the barrel vault; the six central caissons are absolutely regular compared to the width, but the perpendicular lines that define them do not all converge exactly to the convergence point (V); the abacuses of the posterior columns do not fit accurately to a regular plan and seem projected incorrectly. Then Kemp concludes his studies deducing that the Trinity's construction is not consistent and regular in every detail and that any construction

[4]About thu use of perspective and about the space: G. J. Kern, Das Dreifälttigkeits Fresko von S. Maria Novella, in "Jahrbuch der Königlich Preussichen Kustsammlungen", 1913.

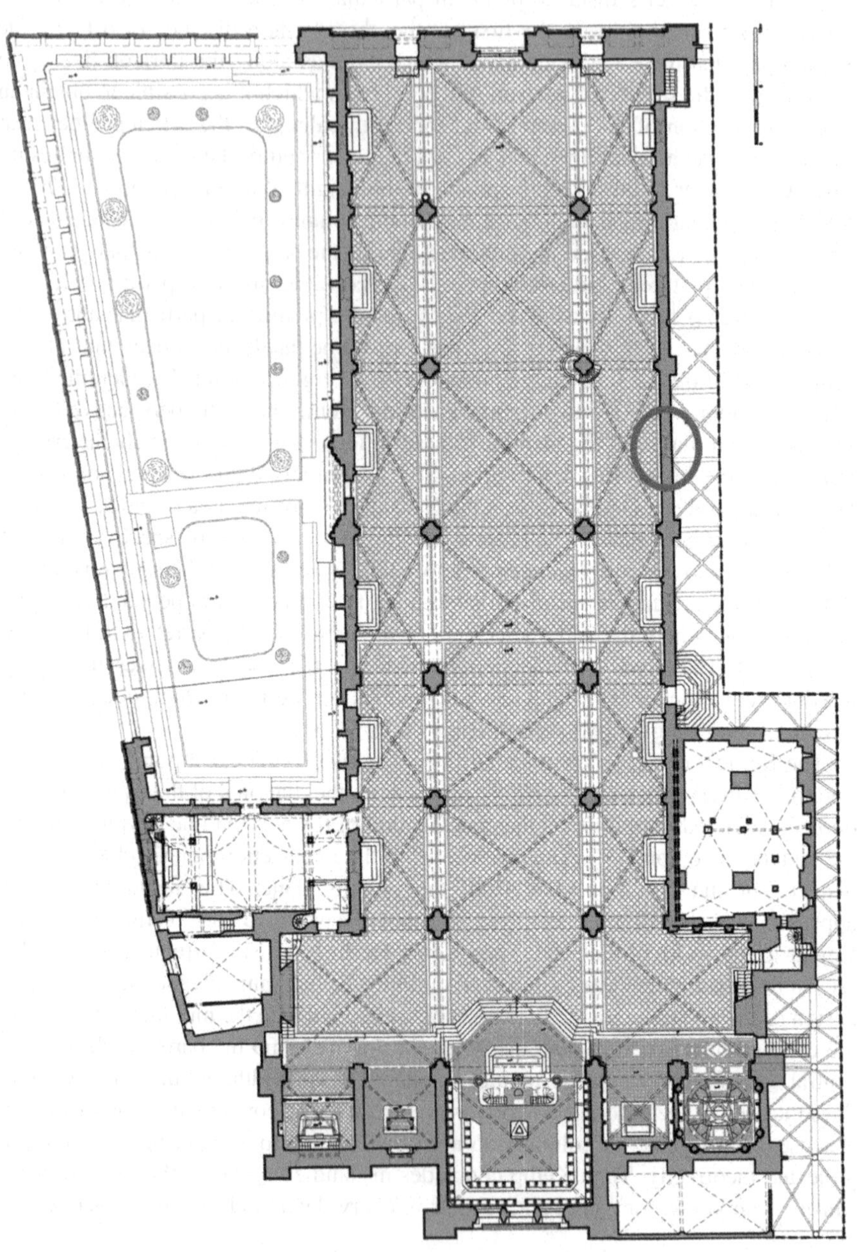

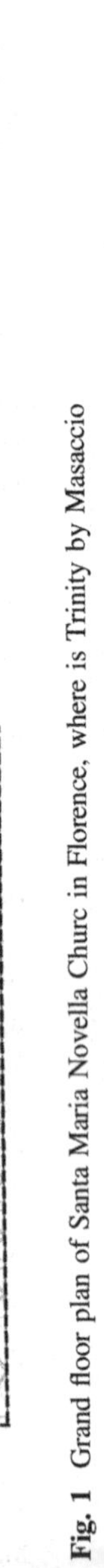

Fig. 1 Grand floor plan of Santa Maria Novella Churc in Florence, where is Trinity by Masaccio

Fig. 2 Perspective drawing theorize by M. T. Bartoli, overlapping picture

can solve entirely his scheme. In some recent studies that deal with the perspective reconstruction of the painting, also Camerota's study produces some theory to think about. The observer's height, as he writes, is about 1.70 m. The convergence point is located just up the altar table. There is approximation about the identification of the observer's distance, with hypotheses ranging from 2,105 m by Polzer (1971) to 8,942 m by Battisti/Degl'Innocenti (1976)[5]. It is not visible the plane of the floor, therefore we can only analyze the elements in the upper part of the painting to define the distance of the point of view. As in previous analyses, Camerota recalls the need to find a square in perspective whose diagonal extended to the horizon line (line through the central of convergence point) indicates the viewing distance (Figs. 4, 5, 6).

[5]For the different results obtained in the evaluation of the viewing distance, cfr.Kern 1913 (524 cm); Sanpaolesi 1962 (572 cm); Janson 1967 (700 cm); Polzer1967 (210,5 cm); Battisti-Degl'Innocenti 1976 (894,2 cm); Myers 1978(729,5 cm); Hertlein 1979 (552 cm); Angeli-Zini 1980 (531 cm); Pacciani 1980(614 cm); Field-Settle-Lunardi 1989 (689,25 cm); Huber 1990 (333 cm); Kemp1990 (540 cm); Kuhn 1990 (642 cm); Aiken 1995 (219,9 cm); Hoffmann 1996(452 cm).

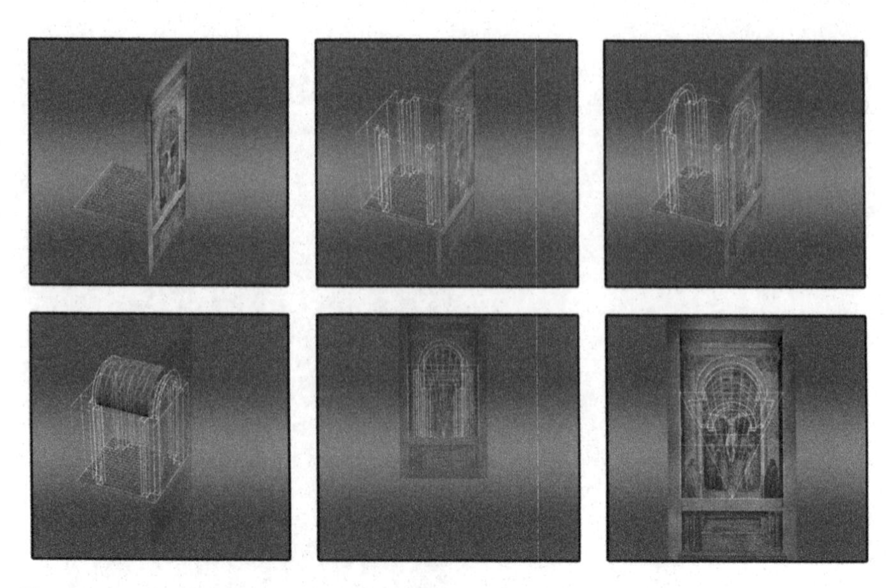

Fig. 3 Animation sequence inserted in the video on the Trinity by Masaccio

Fig. 4 3D model realized for the prototype of augmented reality

Analysis conducted by J.V. Field, T. Settle and R. Lunardi about the square most immediately identifiable, recognizable through the ionic capital's abacus lead, the authors to conclude that the side in perspective of the abacus was deliberately stretched for reasons, it is assumed, of composition: the diagonal that is obtained, therefore, does not lead to reliable results. The focus then moves to the ceiling lacunar. Also in Camerota we can read that these lacunar, in analogy with the architectural models of the time, must be square and equal. This hypothesis appears, however, questionable to author, as the lacunar immediately above the floor of start

Fig. 5 Image frame of sequence in augmented reality

are higher than the others. The transverse ribs are not uniform: the first two would be slightly raised to hide the entire first front of the ceiling coffer.

Therefore, Camerota not theorize a distance point on the basis of the vault's design. He believes, therefore, acceptable that into the picture there may be some adjustments, needed to improve all result and that it is "virtually" impossible to find a picture without projective defects, also when the rule is applied rigorously[6]. Therefore, Masaccio wanted to represent a vault with square ceiling coffe. Drawing the longitudinal section of the picture, it draws the line profile of the horizon and the line connecting the last rib with its corresponding prospective on the plane. In this way Camerota identifies a point away at 6.67 m, a value very similar rispect that already postulated by J.V. Field adds that as this measure corresponds to the width of the nave where the picture is located and also equal to the height of the picture, as if there was correspondence between the observer distance and size of the picture. This generates a viewing angle of 54°, a value that is also found in Brunelleschi and its baptistery's tablet. In the studies cited, however, it is unable to provide a prospective outline of the picture that does not consider a possible approximation adopted by the painter. It ignores, therefore, whether there might be a projective

[6]Alberti 1973 e 1980, lib II, capp. 31–32: "Non credo io al pittore si richieggainfinita fatica ma bene s'aspetti pittura quale sia rilevata e somigliata a chi ella siritrae".

Fig. 6 Image frame of sequence in augmented reality

rule and transported on the picture as it appears to us, considering the fact that the studies analyzed, especially those made before the restoration of the picture, were based on reproductions few conformity to the original, with proportions almost always different. It is therefore understandable that it was not easy to lead plausible deductions. The evolution of digital photography and image sampling process allowed the subsequent studies of using superior quality material reaching, thus, more reliable results. At this point, as writes M.T. Bartoli, the challenge has been the definition of a hypothesis of accomplished architectural scheme that might be consistent with the perspective image, demonstrating that the picture does not show errors or inaccuracies, but it is based on an project similar with architecture structure and construction perspective. The graphic representation proposal responds to this challenge and demonstrates that in the picture all the architectural signs are well resolved and that there isn't any approximation as previous studies suggest.

The study based on the perspective reconstruction, has allowed us to demonstrate that Masaccio's painting has no errors or approximations, rather it is based on a project of harmonious architectural structure and perspective construction. In this specific case the animation has better allowed to translate the three-dimensional model's painting in a new and much popular language, suggesting an innovative approach to its understanding. The animated diagrams intend to show the logical processes highlighted by the perspective interpretation, in an understandable way even to those who are not familiar with the graphic processes of perspective. Its purpose is to show the own real potential of expression by searching the conceptual dynamism inside the apparent Renaissance "static" nature.

This kind of video has to be structured with a "direction" which coordinates images, animation and sound, so that the entire product is not trivial. It must have the ability to capture the viewer's attention and bring him to reflect on the content of the represented object. An important role is played by the sound. It's crucial not only the choice of appropriate soundtracks, but also the presence of a narrator can help to understand the video itself. The video on the Masaccio's Trinity [7] has been structured so that the animation plays a main role. At the beginning there is the description of the graphic construction process of a cube that allow to lay the foundation for understanding the perspective drawing of Masaccio. Afterwards the attention focuses on painting: 3D animation shows thus the wireframe [8] construction of the structure departing from the grounds and going up to the elevations, describing in sequence the performed steps for reaching the final conclusion. The entire work is accompanied by a soundtrack and a narrator.

This animation lays the goundworks for another experiment, launching a project of "augmented reality" thanks to the collaboration with a "mobile developer"[9] for the construction of an experimental app. In order to obtain a product which was not a mere 3D display, it has been necessary to provide a three-dimensional model as far as possible in line with the painting and especially based on prospective studies carried out. The final work is increasingly developing, but it has been created a prototype application for smartphones which follows the above mentioned and analyzed principles.

[7]The study dealt with the Trinity adopted modeling software, rendering andediting in post production. The programs used were: 4 Rhinoceros for modeling,3D Studio Max for mechanical animation, V-Ray for 3D Studio Max for rendering,and Adobe Premiere CS5, and Final Cut Pro X video editing and text.

[8]wireframe: model in wire, a type of graphic representation of three-dimensionalobjects. With this method are drawn only the edges of the object which remainstransparent in its interior, as if it was built with the "wire".

[9]The experiment of augmented reality realized by Doct. Giovanni Landi, MixedReality Architect.

3 Conclusion

The representation of the third dimension on flat surfaces has been a challenge for artists of various ages, ever more linked to the need of combining deception and illusion. The use of animation for transmitting the perspective model's perception, its content and its geometric process, should be the right way to make these topics accessible to a wide audience of users without sacrificing the scientific accuracy. A similar approach to the subject can also tempt the non-experts of the field to approach these studies, integrating the more familiar interest in the historical and artistic aspects, with the geometric-composition. In this way it's possible to approach the understanding of the perspective architecture system which is able to affect the viewer going over and expanding the real space, as in the case of Masaccio'sTrinity

References

1. Jakobson, R. (2002). *Saggi di linguistica generale*. Milano: Feltrinelli.
2. Dalai Emiliani, M. (1980). *La prospettiva rinascimentale*. Firenze: Centro Di.
3. Kemp, M. (1990). *The science of art: Optical themes in western art from Brunelleschi to Seurat*. New Haven: Yale University Press.
4. Camerota, F. (2006). *La prospettiva del rinascimento Arte Architettura Scienza*. Milano: Electa.
5. Bartoli, M. T. (2014). *Brunelleschi e l'invenzione della prospettiva, in Prospettive architettoniche conservazione digitale, divulgazione e studio*. Roma: Sapienza Università Editrice.

Research of Nonlinear Architecture Morphology Based on Psychology Rules and Mathematical Orders

Jingting Cheng, Jianqun Lin and Weiming Zhang

Abstract Technological development is to serve the needs of the human body and mind. Today's digital technology has become a relatively perfect platform, it has built a new generative architecture form, with numerical simulation, biological information, natural morphology and mathematical rules. Furthermore, we want to know the suitability of such a strong impact on people's physical and mental needs or aesthetic rules. This paper is trying to explain and solve this problem. First step, use examples to find out the morphological features of non-linear architecture which are different from the traditional architecture. Coming out four results: ambiguity of space, temporality of direction, uniqueness of position, and information of volume. Second step, according to the four results of argumentation, seek for solutions. The solutions are based on psychological rules and mathematical orders: setting a visual gravity point, using light to find the direction, invisible axis, and "blank-leaving" of scales, finally draw a conclusion. Hoping the research will provide theoretical guidance of parametric nonlinear architectural design in the future. The purpose is to avoid technological breakthroughs to bring psychological pressure and aesthetic missing.

Keywords Nonlinear architecture · Architecture morphology · Psychology rules · Mathematical orders

J. Cheng (✉) · J. Lin · W. Zhang
School of Architecture, Harbin Institute of Technology (HIT), Harbin, China
e-mail: 11b934002@hit.edu.cn

J. Lin
e-mail: linjianqun50@163.com

W. Zhang
e-mail: zwm516@163.com

M. Rossi and G. Buratti (eds.), *Computational Morphologies*,
https://doi.org/10.1007/978-3-319-60919-5_17

1 Introduction

The architectural morphology is the complementary infiltration between "shape" and "status" of architecture. Compares with the shape and form, this research on morphology is focus on grasp the expression of architecture, namely, the dynamic effects and the psychological effects [1]. "Shape" constructs the architecture from outside to inside, from static to dynamic. "Status" expresses the architecture from the inside to outside, from dynamic to static.

The analysis on Gestalt psychology "sentient" by Arnheim R., concluded that the visual style is actually a field of strength. Assuming these "strength" as some as real power, existing in mental space and also in the physical world. Biology believes that the brain's visual cortex is an electrochemical power field, Wei Taimo (1880–1943) once operated an experiment on the feeling of movement: in a dark room, he put two light spots in different locations, let them shining successively within a very short time. The observers report that they are not seeing the two spots of light separated from each other, but there is a with movement from one spot to another. It calls physiological short circuit, the energy moves from one spot to spot.

That means, in the cerebral cortex, the interaction between local irritation and local stimulation spot is the interaction of power [2].

Architecture morphology brings people in or around the architecture produces a particular feeling, affect and function of the environment from architecture morphology called the morphogenetic field [3].

In the process of people perceiving architecture morphology, the strength field of psychology is the architecture morphology field. Actually, architecture morphology is the dynamic equilibrium after through the constitute interaction of the various strength, mutual support, mutual offset of the above.

The balance will change with the ambience, the time climate, the observe order and other factors change, and then, people are instinctively to adapt to these changes. In the four-dimensional digital architecture space, time is the ruler of changes, time is able to describe the changes. Non-linear architecture morphology use the time dimension to become dynamic equilibrium of all the changes. The relationship of strength in morphology, can give life vitality and timeliness to architecture, which is particularly important to parametric nonlinear architecture.

2 The Research

2.1 Analysis

The basic elements of architecture morphology field are space, direction, location, volume and so on. Parametric architecture broke the three-dimensional space of traditional building construction, "curve" became the aesthetic expression of digital design, algorithm, generation, emerge, cooperation became the most popular keywords [4]. The nonlinear architecture based on these design concepts, its morphogenetic field shows the following features.

2.1.1 Ambiguity of Space

Architecture morphology must be attached to a certain space to exist, and produce its aesthetic value.

From a psychological perspective, the architecture morphology we are perceived is the main part stand out from the deep space, in turn, the space is perceived as a minor part of the following topics from comparison to the main part.

However, unlike the previous traditional architecture, nonlinear construction, its prospects and the background is dissolved in random expression.

In or out the nonlinear architecture, anywhere could be central part or may be background space, the space and the morphology are always transforming to each other (Fig. 1).

Spatial properties have decisive significance to architecture constitute, borderless space led to the disappearance of the "field". We already have the developed aesthetic experience, it depends on the relatively limited spatial extent, cognitive and control the size of the whole lack of spatial relationships, orientation and other factors.

When the nonlinear architecture providing us an unlimited space, it also get us easy lost, we may not find the center of gravity (Fig. 2).

Fig. 1 Internal open space of the Perfumed Jungle building

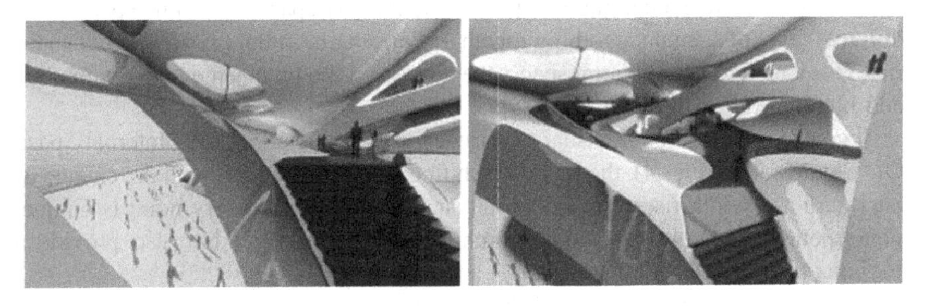

Fig. 2 Interior space of Nuragic and Contemporary Art Museum

Fig. 3 Rome National Museum of the XXI Century Arts, by Zaha Hadid

2.1.2 Temporality of Direction

Rome MAXXI (National Museum of the XXI Century Arts) is a complex sequence of continuous ultra-performance practices of space work designed by Zaha Hadid. It uses different elevations rectangular blocks composing the Art Center exhibition hall, to reflects the space's timeliness (Fig. 3).

Each exhibition space communicates with the atrium, so that visitors can feel in a different way anywhere in the communicates, or with the penetration level on each space between cells.

The arrangement of streamlined channel connected the various elevation, with activities to strengthen the liquidity and continuity of continuous space.

It looks like extension of time, so that people feel the presence of multiple levels of space and complex at the same time, ultra-order temporal experience.

In that kind of space, the time dimension interacted with each visual direction, so it has a temporal direction, time made the direction weakly.

2.1.3 The Uniqueness of the Position

Interior design of nonlinear architecture also reflects the complex sequence space (Fig. 4), it looks like a continuous spiral, functionality space and traffic space, with different level of spatial smoothing together. Forming morphology simple but super rich visual experience program space, the space and the time achieve to continuous and blend in different dimensions [5].

It gives visitors different visual experience in every position. Like Internal Space Of "Sky Mound" (Fig. 4), the link of each floor walkway ladder, with the change of curvature and width rises upward spiral, such a space walk, every point is unique, non-repetitive, it raised a complex architecture morphology, but also provide a tortuous path to aesthetic psychology.

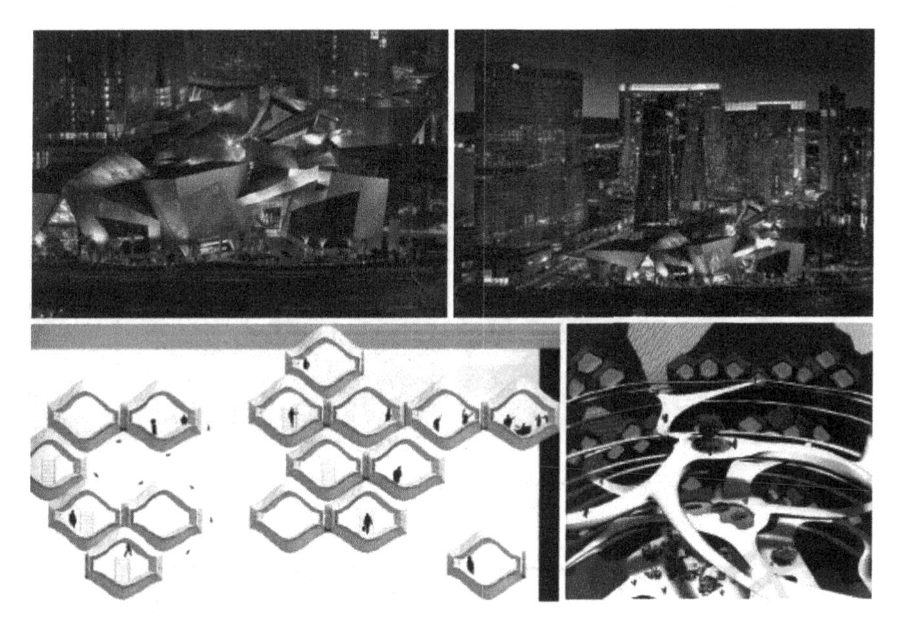

Fig. 4 "Time Crystal", Libeskind *(above)* and Ant Nest Structure of "Sky Mound" *(below)*

2.1.4 Information of Volume

In Eden Project, in order to highlight the "biome" design theme, the building is a "cell texture" simulation, lead to a hexagonal and pentagonal modular steel frame structure. This kind of morphology, using amplifying expression of onlookers nonlinear shape to give the building a biological skin style, delicate, system [6]. They put biological information, the natural model, bionic principle inject to the architecture design program. It is undoubtedly a very successful attempt, these forms are very easy to resonate, but the density control is tricky, parametric generation transformed the volume of architecture into information. In fact, what we see are lots of information node, the node is likely to cause an overload of information for those who experience psychological pressure (Fig. 5).

2.2 Discovery

"Mathematical order" in the title is the outcome of these highly abstract classification, integration, statistics from the laws. The book "Creating Orders" from Chinese Dynasty Song, summarized the study of Chinese architecture modulus systems theory and harmony (Fig. 6).

Chinese architecture has a decisive influence on a group of Western architects. Danish architect Utzon carrying "Creating Orders" and careful study, during the

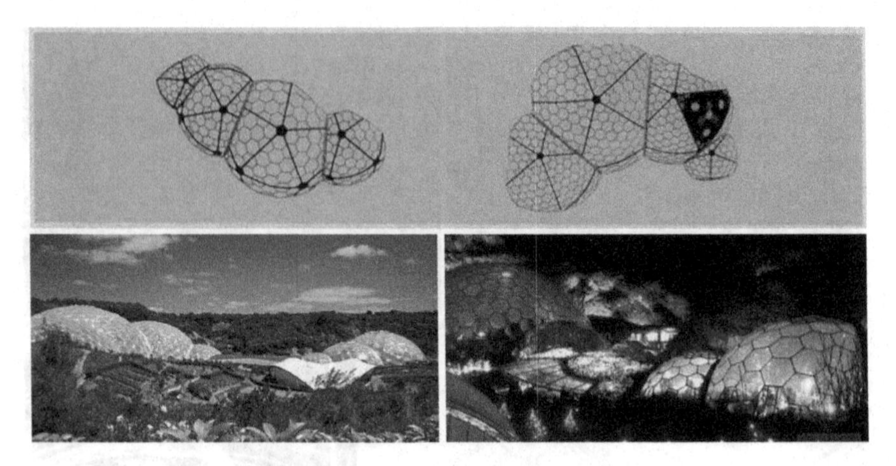

Fig. 5 Structure and outward appearance of "Eden Project"

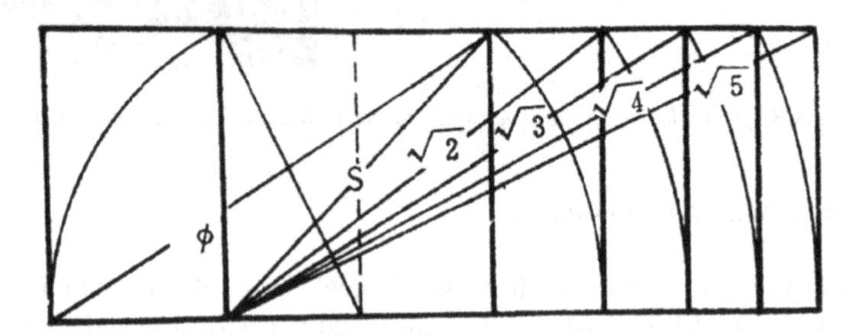

Fig. 6 Rectangular proportionate consolidation chart

Sydney Opera House program, Utzon used the idea of "Creating Orders" theory to solve the design and structure problems [7]. In the design process, to build a "movement in stable morphology" is the task of architects. In fact, the most simple psychological rules and mathematical orders is the best solution to solve these problems (Fig. 7).

2.2.1 Setting a Visual Gravity Spot: The Golden Ratio

If you want to strike a balance between psychological force, you need to find a visual gravity spot, the easiest way is to find the most suitable ratio between the environment within a particular area of architectural morphology. Such as the ancient Greek Parthenon Pa bottom. Static balance is too rigid, people are more inclined to dynamic balance, finding a sense of order and a coordination aesthetic feeling. So people found the golden ratio, scientific proofed, it is the aesthetic experience of building the most comfortable form of gravity point (Fig. 8).

Fig. 7 Arab World Institute and Mobile Phone Experience Center

Fig. 8 Nonlinear light orders

2.2.2 Using Light to Express the Time: The Dramatic Conflict

Time exists in the speed of light, the light on the building are constantly changing every moment. The change follows the same rules, with light in the direction of non-linear architecture specification, just like a movie. There is before and after, and dramatic conflict of transforms. In the future, we will rely on light to express the temporal dimension [8]. For example, many optical architecture of modern art in the Light “live” art space, outside the “internal” work, and allow the exchanging between “past” and “now”. We walk in the complex non-linear building with time-space, and to find the loop through light (Fig. 9).

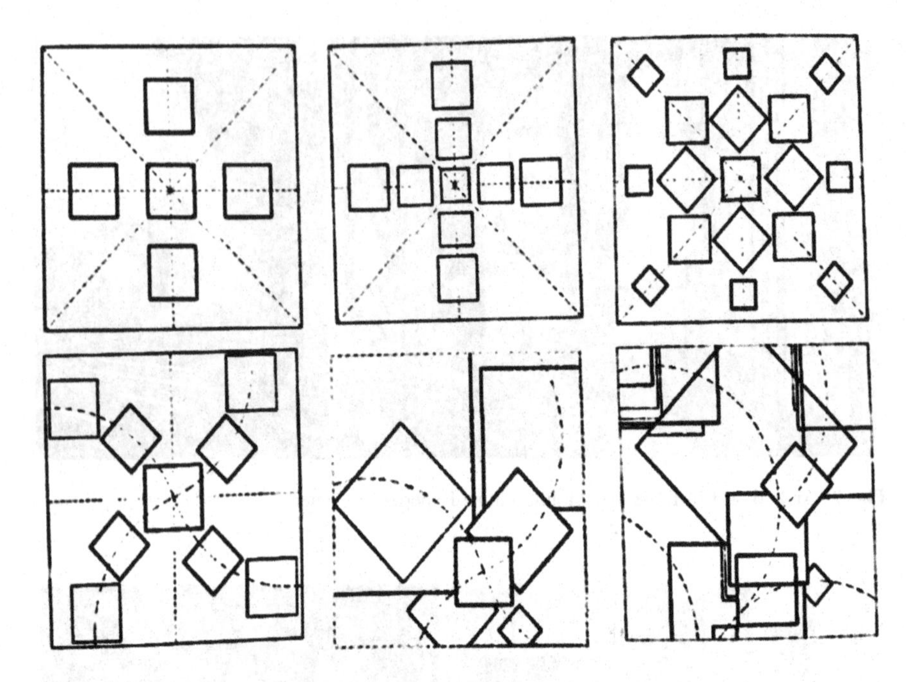

Fig. 9 Hidden linear axis and hidden curve axis

2.2.3 Layout Hidden Axis, Invisible Balance

Sense of order is the important performance of architecture's coordination, integrity, aesthetic, its core is requiring the relationship between the various elements associated with a regular. Some are obvious, some are hidden invisible, which including various forms of relationship and dynamic law created in space power. Perhaps, axis is the best way to express the sense of order, it has a powerful ability to dominate and control the various elements of the arrangement, which should focus on the specific location of the axis determines around axis, the architecture morphology is uncertain or overriding, the structure is loose or clarity, activities vivid or tedious [9]. Most of traditional architecture's rhythm expression is in appliances, in the element surface. For example, you can use the same columns on traditional building, spaced them in same distance, in order to express the rhythm, or use the same form of relief at key nodes of architectural detail, to express rhythm. Parametric nonlinear building would need to put their orders "invisible" to the hidden psychological power among morphology rules [10] (Fig. 10).

2.2.4 Blank-Leaving Scales: Highlights Tension

When two or more visual images in people's view, even when separated by a certain distance between them, there will be a visual structure tendency. Some

Fig. 10 Powerful balance forms

relationships quite calm, some in the contest among; sometimes delicate situation, in the contest with the intermediate state of harmony, making full of tension [11]. Elegant mathematical order has quietly disappeared in our present system modulus. In fact, it still should be returned to the non-linear construction, and in a new way [12]. Classical physics believes that all objects have the quality to each other there is a gravitational power, and this force is proportional to the size and quality of the object. Although this attraction is not exactly the same as a physical strength, we can borrow it to analyze the physical principles. In general, the attractiveness of their properties related to images, such as the size of the relationship between the maximum contrast images attract the strongest relationships, contrast is smaller, attract relationship weaker. Therefore, the to build a "powerful" non-linear architecture, the key is not complex structure, and this tension is necessary scale changes to get that blank-leaving scales.

3 Conclusion

We know that people's perception itself has a sporty, nonlinear structure using these sporty, these works usually give people a strong visual stimulation, brings powerful effects. This perception is the psychological strength generated in the contact process, the architects used abstract geometric forms, exaggerated body mass scale, continuous motion curves, strong color contrast and other techniques, to break the traditional architectural spatial order. This is positive in the beginning, and effective, but now, under the rapid development of digital technology, in some cases, excessive pursuit of temporal distortion has exceeded people's psychological acceptance. We live in the real space with cycle laws, the advantages of non-linear architecture is unlimited, but should not without order. The order can be found in traditional mathematic, can be found from the mechanical psychology. Using psychology rules and mathematic orders to design nonlinear architecture, we can bring architectural morphology resonance. After all, architectural aesthetic standards is not gorgeous and unreal, but the balance and harmony.

References

1. Lin, J. (2000). *Modeling foundation*. Beijing: Higher Education Press.
2. Arnheim, R. (1998). *Art and visual perception*. Chengdu: Sichuan People Press.
3. Chen, Z., & Chen, Y. (2006). *Architecture morphology*. Beijing: China Architecture and Building Press.
4. Leach, N., & Yuan, F. (2011). *Scripting the future*. Beijing: China Architecture and Building Press.
5. Liu, Y. (2013). *Research on contemporary architectural thought under Deleuze's philosophy*. Doctor Degree Thesis of Harbin Institute of Technology, Harbin.
6. Zhang, M., & Liu, S. (2012). *A study on the aesthetic characteristics of form diversity in contemporary western architectural symbols*. Harbin: Urbanism and Architecture.
7. Li, Y., Fang, H. (2005). Human, measure scale and proportion. Finnish Master Architect Blomstedt and Pythagoras Harmony Tradition. *Huazhong Architecture* (Changsha)
8. Tanca, M., Grossberg, S., & Pinna, B. (2009). *Probing perceptual antinomies with the watercolor illusion and explaining how the brain resolves them, seeing and perceiving*. Leiden.
9. Liva, G. (2008). Section and projection: The cinematic experience of the conical sections in Antony McCall's work. In: *Descriptive geometry and digital representation: Memory and innovation*. Milano.
10. Ching, F. D. K. (2008). *Architecture: Form space & order*. Tianjin: Tianjin University Press.
11. An, H., & Gao, X. (2013). *Analysis on visual dynamic type of architectural form: Based on the visual perception theory of gestalt psychology*., Architecture & Culture Beijing: World Book Publishing Company.
12. Caldas, L. (2008). Generation of energy efficient architecture solutions applying GENE_ARCH: An evolution-based generative design system. *Advanced Engineering Informatics, 22,* 59–70.

Generating the Pantheon's Dome: Cultural Paradigms and Shape Grammars

Nevena Radojevic

Abstract The paper examines some of the possible ways to generate the coffered domes, where generative grammars are determined by different cultural paradigms. A method applied is the shape generation method, that uses shape grammars, which take shapes as primitive and have shape specific rules, as described by G. Stiny and J. Gips. The generation of the three dimensional shape consists in the definition of a language of two dimensional shapes, the selection of the shape in that language, a specification of a sphere mapping schema and the determination of a location and scale of the shape on the series of concentric spheres of a given size [1, 2]. This work shows that the essential differences between the Pantheon's dome and some "Pantheon inspired" coffered domes are caused by applying different generative grammars, characterized by their cultures.

Keywords Pantheon · Shape grammar · Generative design · Stereographic projection · Coffered dome

1 Introduction

What we can define as "generative design" is a design that it is not guided by a sequence or a database of events or forms, but, by a definition of a set of behavior patterns. The design settles a set of transformations rather than a form, and the form is just one of the possible results from the process. So, the emphasis of the creative process moves from the *form finding* onto the definition of *transformation rules* [3–5].

If we try to look for the rules that have been applied to generate some of emblematic examples of the coffered domes, we'll be actually trying to define a kind of an "inner language" that belongs to that structures, that is their shape grammars. Once we have deciphered the language, the same set of rules could be used to create the same "species of architecture" (Soddu [6]), or as an instrument for

N. Radojevic (✉)
Department of Architecture, University of Ferrara, Ferrara, Italy
e-mail: nevena.radojevic@gmail.com

M. Rossi and G. Buratti (eds.), *Computational Morphologies*,
https://doi.org/10.1007/978-3-319-60919-5_18

their recognition, as they follow the same logical procedures and the same design approach. The examples on which we are going to research the coffers generation methods are: Rome's Pantheon (classical antiquity) and the church of the Abbey of St. Genevieve, or the so-called Paris' Pantheon (neoclassical period). In this paper, we will examine some of the possible ways to generate the lacunar domes, with emphasis on the shape grammar that produced their final forms (Figs. 1, 2).

2 Some Methodological Premises

The research, in the case of the Pantheon in Rome, is based on the metric data provided by integration of few surveying campaigns, and, in part, is founded on the previous studies.[1] The metric sources of this study are:

- Three dimensional laser scanner data (cloud of points) that has been made available to us thanks to the University of Rome and prof. Riccardo Migliari's research group. This data consists of 12 separate scans of the cupola, without a topographic base [10].
- Snapshots obtained from point cloud of a complete Pantheon, processed by University of Bern, and available on an online platform.[2] The Bern's University data was integrated with the first one (the metric one), in order to verify the registration process and cover the missing zones.[3]

As we mentioned, the cloud of points has been acquired without a topographic network support and with a very tight field of grip, which has produced 12 different scans, not oriented in respect to the horizontal plane. For these reasons, it was difficult and not very reliable to do the recording based on the physical points. Therefore, we preferred to process each separate slice (Fig. 4). The first stage of processing has been the research of a sphere that best approximates the inner dome's shape, and the examination of the surfaces belonging to successive concentric spheres. With the aid of computer tools (Rhinoceros and Grasshopper) and some scripts made for the calculation of the adhesion between the points of the cloud and the points on hypothesized sphere, we are able to make some observations: the diameter of the best approximating sphere of all the points of the intrados is 144 (122) roman feet. The 30% of the points of the cloud has the greater distance

[1]The Pantheon section's shape generation study is based on a previous thesis made by prof. M. T. Bartoli that hypothesized the possibility that the dome's design was made in according to the use of Ptolemy's stereographic projection, putting it in a close relation with Vitruvius' scenography [7–9]. In that period, being the digital surveying data not accessible, it was very difficult to describe the section's profile in a precise manner. The further studies and verification were carried by the author in part in Ph.D. Thesis, and in part for this occasion.

[2]The Bern Digital Pantheon Model, http://www.digitalpantheon.ch/building/digitalmodel.

[3]I am using this occasion to thank them for sharing the survey data with the scholarly community and encourage the research that way.

of 15 cm from the ideal sphere. The points that deviate more are near the oculus. Considering points near the oculus the result is a sphere of a diameter of approximately 146 feet. So, there is a slight flattening of the dome in the upper part, caused by its weight.[4]

In the case of the Pantheon in Paris, the thesis is based on a treatise written by the architect who had completed the dome, Giovanni Rondelet.

3 The Coffered Domes

The coffers are cavities formed in the ceilings of the spherical, cylindrical or oval domes, arranged on a regular basis. They are characterized by the geometric shapes, among which the most common are: the circle, the square, the rectangle and the octagon. For this reason, the lacunar's design is commonly linked with the scientific reflections on the issues of the sphere and its representation. It represents the necessary base for the celestial and terrestrial studies (astronomy and geography). The different ways to track the lacunars onto the sphere surface, often reveals a particular set of cultural paradigms to which they belong.

Giovanni Rondelet, executor of the coffered vault of St. Genevieve church (also known as Paris's Pantheon), designed by Soufflot, in one note of his famous treatise entitled "On the way to describe the coffers on spherical and spheroidal vaults",[5] explains the method he used for the realization of various lacunar types (Fig. 3). According to Rondelet, the design of the coffers on the spherical vaults is solved once we have found the rays of the series of tangent circles that can be plotted, inside the unrolled spindles, between two ribs and the horizontal circle. Once the spindles are developed, using the approximate method that he described in the chapter titled "Sviluppo dei solidi la cui superficie è a doppia curvatura", we are able to draw the first circle tangent to the first horizontal arch and two vertical arches, the next is tangent to the first one and the two vertical arches, and so on. Subsequently, we can inscribe any other regular polygon inside the circle. Once we have chosen the polygon's shape, we can draw a series of concentric polygons on the developed spindle and then generate the section (by transporting the distances). The three dimensional form is created by assigning the extrusion quantity of the shapes, and the side edges between the adjacent squares (for example) are perpendicular to the spindles plane, thus, oriented to the sphere center [11].

[4]The similar result was also found by prof. Graziano Valenti who had examined the same surveying data, recorded in one unique cloud of points that time, and the radius value found by him ranged from 21:40 to 21.90 m. So we can consider this values as verified.

[5]Giovanni Rondelet, Trattato teorico e pratico dell'arte di edificare, Mantova 1832, tomo II, parte I, "*Nota*" pg. 190–199.

Fig. 1 Pantheon in Rome. Photo by author

4 Pantheon Dome's Design

When we look to a vault designed in the "Rondelet's" manner, although it was designed by applying the sequence of concentric squares rule, we cannot observe them when we look at the dome. Some of the side edges are not visible any more, and some of them are aligned with the visual rays. On the other hand, if we observe the Pantheon's dome from the center of the ground-floor, we will see a completely different image. We would see the 28×5 series of concentric squares, where all the edges are clearly visible, including the side ones, as if the three-dimensional design was a completely flat drawing projected on the inner sphere. The coffers are visible as a sequence of concentric squares, and we cannot even read their 3rd dimension without the presence of light and shadows. As C. Soddu noticed: "… before the light you don't understand if the sequence of frames is going toward the inside or toward the outside of the sphere. …. And something incredible can happen. With the sunlight the 3d objects reverse their position. For one moment you can understand the Pantheon as if it was in a Florenskji's reverse perspective, and all you are seeing in the inside is the outside skin".[6]

On the other hand, the curious vertical section of lacunar appears as very chaotic: while the edges of the upper parts of the profiles converge towards the center of the sphere, the lower ones have a completely irregular convergence. This very particular

[6]C.Soddu, E.Colabella, "Lumen in Splendore", proceedings of ApliMat, University of Bratislava, 2008.

occurrence which does not seem to happen on the other Pantheon inspired domes, is led by very specific and sophisticated design pattern. In order to decipher that pattern, we need to ask ourselves the following questions: Which is the rule that guides the degradation and the distribution of the squares in 3rd dimension? How is it possible to see the squares as if they were concentric while they are placed on the four different spheres? Which reasons and what kind of knowledge could lead to such a sophisticated project and its execution?

The great visual effect which transmits the image of the Pantheon's dome is achieved by setting the design process in a very sophisticated manner. Such result is only possible if we start with a flat, two-dimensional design (the final result that we want to see), and make a sort of a solid projection on a series of concentric spheres. The solution refers to a particular central projection, the stereographic polar projection, described by Ptolemy in the same period of the Pantheon's construction (or rather re-construction, as we know it today). The polar stereographic projection is a conformal projection, which transforms a plan drawing in a drawing onto the sphere surface and vice versa, while maintaining unchanged angles (a circle on the plan is transformed in a circle on the sphere, a square in spherical square etc.). The center of projection is located in one of the poles of the sphere (the center of the floor in this case), while the projection plane (*quadro*) is given by the equatorial plane (horizontal plane passing through the dome's base). The most interesting step in this process is the way to obtain the lacunar's depth development, which also confirms the method used in the previous phase (Figs. 5, 6).

Fig. 2 Paris Pantheon's dome designed by Soufflot. Photo by author

5 Decrypting the Pantheon's Language

We can try to divide the dome's generation process in three significant steps. The first step, or the "parametric design phase" is the input data phase where we define some parameters. It can be common for all types of coffered domes. The second phase is one in which the process could be divided in two different ways, but the result would be still the same, no matter the choice. And, the third, and most significant one, is the "language definition phase". In this phase we need to answer the questions like: In which manner are we going to generate a 3rd lacunar's dimension? Does its form need to be perceived in a certain manner, or its generation is ruled by some other principles? Or, once we have determined the first sequence of squares are we going to do the simple extrusion of it and then place the others in a concentric way according to their centers, or we want them to be perceived as concentric? The output data, in this phase, is divided in two big groups: the *Pantheon species*, and the *Pantheon-like species*. This phase is able to reveal us also the procedures that have been used in the second one (Figs. 2, 3, 4).

If we look for the possible algorithms, due to describe the generation of coffered dome types, we can define the three main steps (as shown in a diagram):

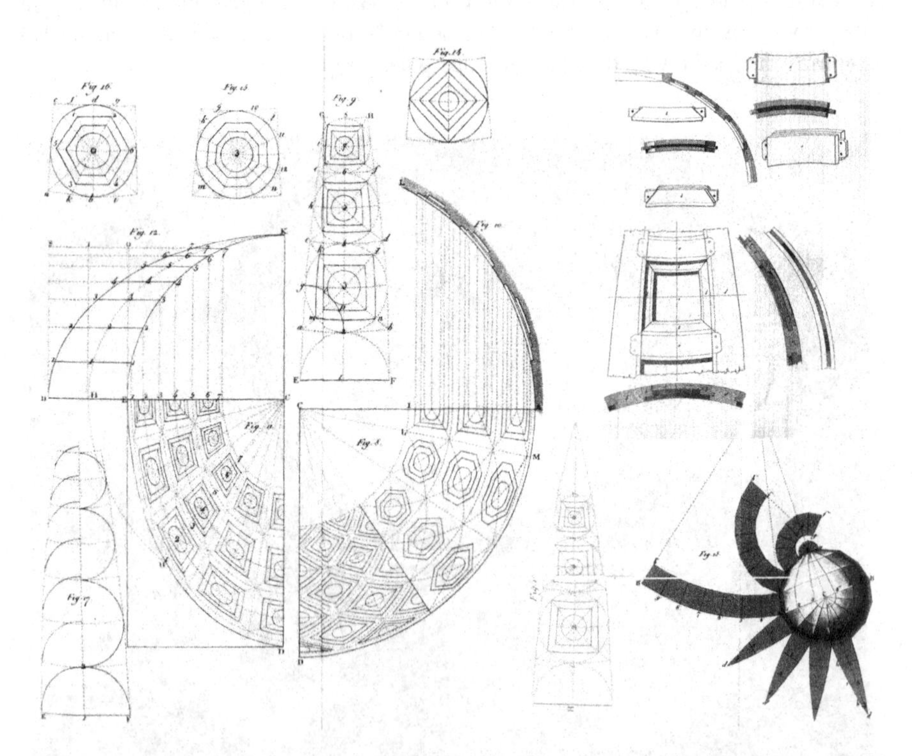

Fig. 3 Assemble from Rondelet's treatise designs: T_II_I/18_XXVI, T_II_I/42_L, T_III II/07_CXXXVII

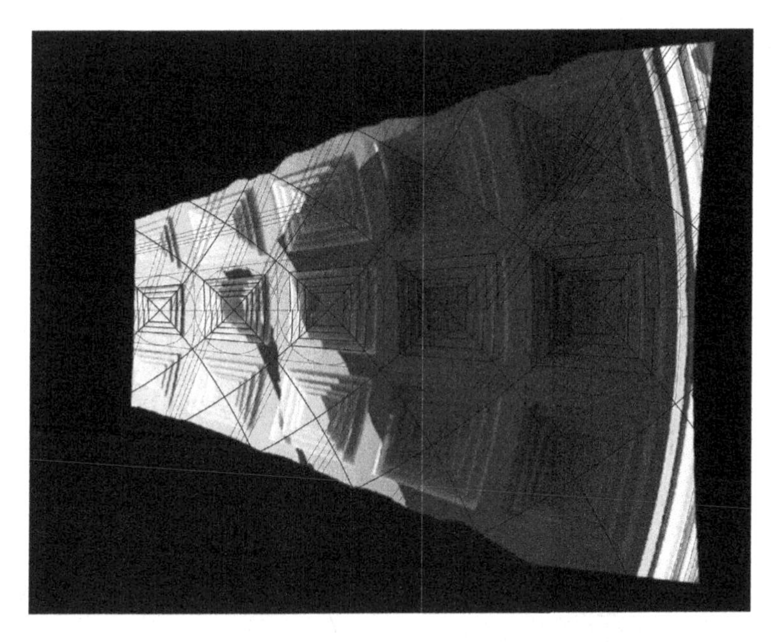

Fig. 4 Mesh model obtained from the point cloud segment—overlapping the hypothesis design

1. Input data set, or *parametric design phase*, in which we are giving the initial input data. The variables in this case are few input parameters, such as: the sphere's radius, number of meridian and parallel divisions and the lacunar's shape choice. These very first steps, even though they influencing the final result, are not revealing the real "language" of the dome that we are creating (Fig. 5).
2. *Mapping the inner sphere phase*, in which we are defining the way to map the dome's inner sphere. As we mentioned, there are two possible ways to map the sphere's surface in a regular manner. The first, and the exact one, would be the stereographic projection (Fig. 5), while the second is given as in Rondelet's spindle unroll method (Fig. 3). Even if the second one is approximate, it's not influencing in a significant manner the final result, so at the end of this phase, we can consider the output data as equal.[7] The design, in the Pantheon's case, starts by dividing the sphere (in stereographic projection) by 28 meridians, and

[7]The geometrical differences produced by these two different processes are insignificant for the structures of this scale.

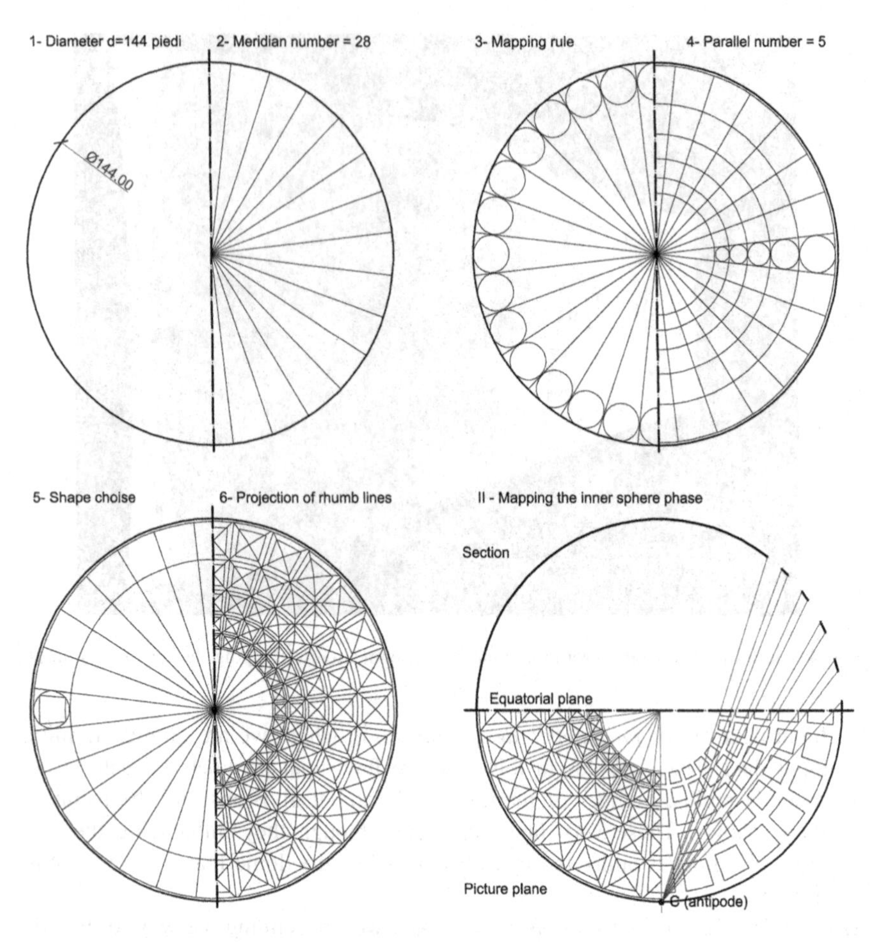

Fig. 5 1st and 2nd Pantheon's design phases

enrolling the first circle tangent to the two meridians and the maximum parallel circle, which determines the next parallel's radius. The next circle is tangent on a new parallel and the two meridians and so on. At this point, we are able to draw the curve that passes through the points given by the meridians and parallels intersections. This curve represents the stereographic projection of the sphere's 45° rhumb line[8] (hence the diagonals of the squares). Attributing the rib thickness between two adjacent meridians also the parallels distance is determined. Now we only need to project them (from the antipode) onto the sphere, and we have found the first series of squares (Fig. 5).

[8]As the result of a stereographic projection of a sphere's rhumb line we have a logarithmic spiral. This design could be also approximated, without any important effect on a final result.

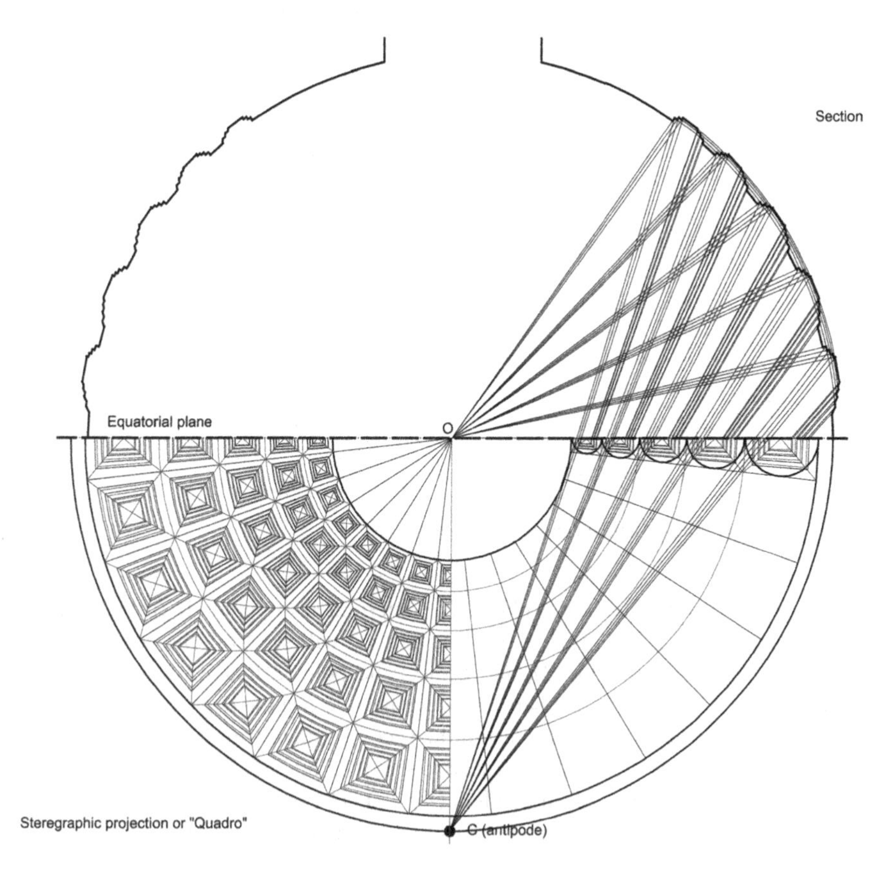

Fig. 6 3rd design phase

3. *Shape grammar phase*, where we are defining the lacunar's 3rd dimension's generation set of rules. Starting with the concentric shape design concept in both cases, the output data, after this step, will be divided in two main groups. In the "Rondelet's manner" we simply need to extrude the sequence of shapes toward the sphere's center, as described in Sect. 3; while in Pantheon's case, the matter is a bit more complex[9] (Figs. 6, 7).

After applying the rule 0, by which we determine the points A′ and B′, we need to project the point A′ from the sphere's center O onto the next sphere's surface, and we got the point 1 (rule 1). Next step is the projection of the point 1, from the

[9]In this step we can choose between two predefined elements: the radius of the next sphere, or the thickness of the side elements that we want to see in their image (perspective or stereographic projection). As both of the procedures are very similar, we will examine only the first one.

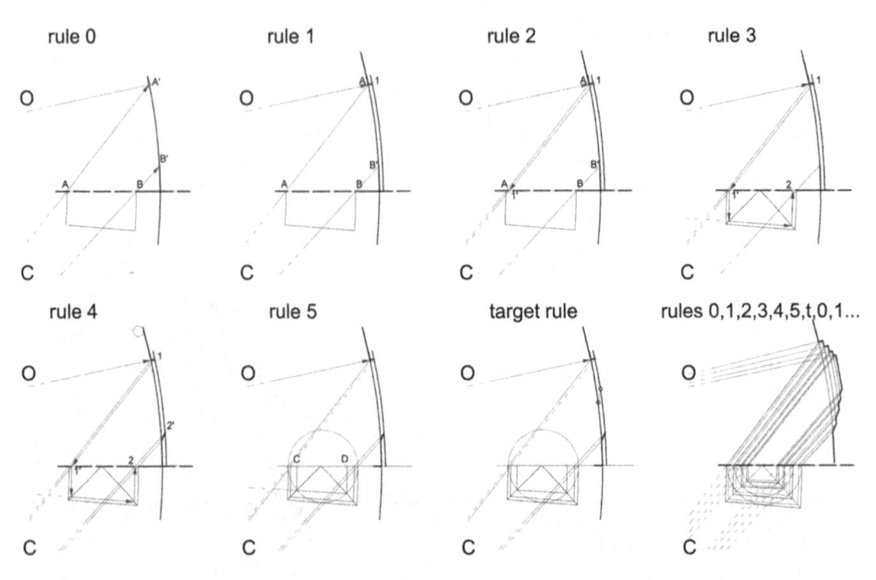

Fig. 7 Shape grammars applied in the 3rd design phase

antipode C, onto the equatorial plane. We determined the point 1' (rule 2). As the point one gives us a thickness of the side that we will see from the center, we can design the next square (rule 3), and project the point 2 that we found, from the antipode C, onto the second sphere. Now we need to use the degradation rule in the stereographic projection, due to design the second square (rule 5). By applying the 5th rule, the target moves from the 1st sphere to the 2nd one. Repeating the same set of rules we can complete all the coffers (Fig. 7). The end of the processes is given by number of repetitions (number of parallels minus 1 in this case) (Figs. 8, 9).

6 Conclusions

The Pantheon's section, obtained in the described manner, is very close to the survey data. Some points adhere almost completely to those of the cloud of points, while others differ slightly. However, considering the sphere deformations caused by weight and earthquakes that the structure has undergone in two millennium period, the result can be considered very satisfactory. The shape of the profile is therefore not chaotic or random as it seemed, but comes out in a very sophisticated and precise setting.

The masterpiece is always a complex artifact, which allows different levels of reading, and multitude of interpretations. It is in itself innovative, both for the content that it provides and in the ways it was made. Any reading (both through survey and historiography) always represents a reverse process, that is subjective and therefore partial and temporary. Each study, however, should at least open a

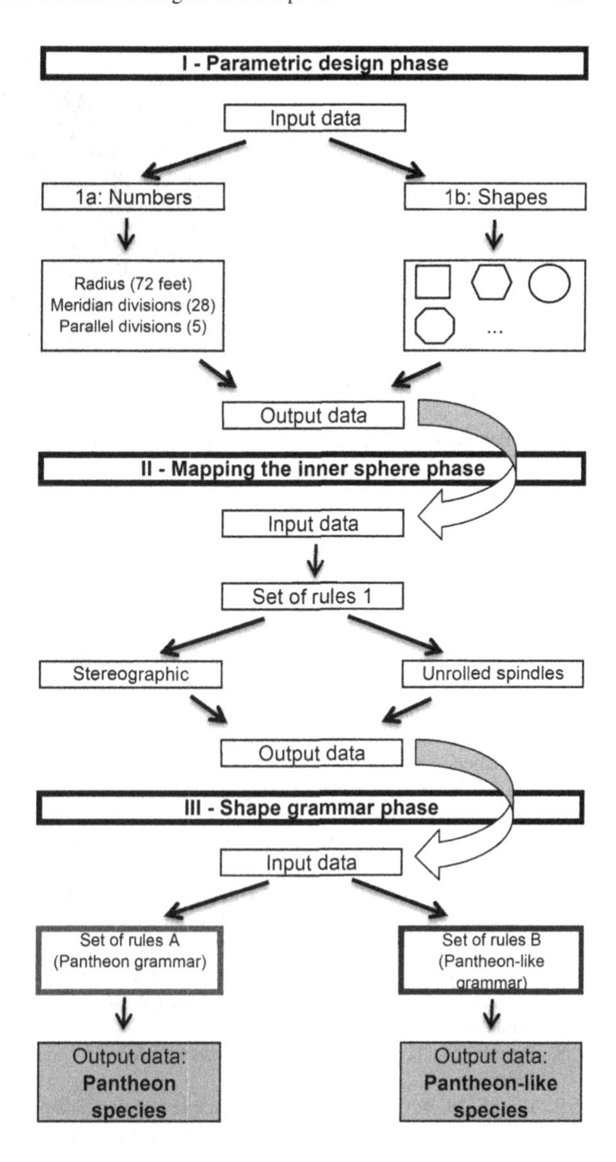

Fig. 8 Schema of possible choices during three generative phases. The only step that is producing two different species is the *set of rules* selection in shape grammar phase

way for further analysis and insights. Influenced by both cultural and design development methods, the "shape grammar phase" can be seen as the key of design process. It's clearly showing how the language of the past, sometimes, can be changed or updated with new technological development and design principles, by the future users. With the same input data but by applying different generation rules, we are able to produce two, conceptually, very different "dome species".

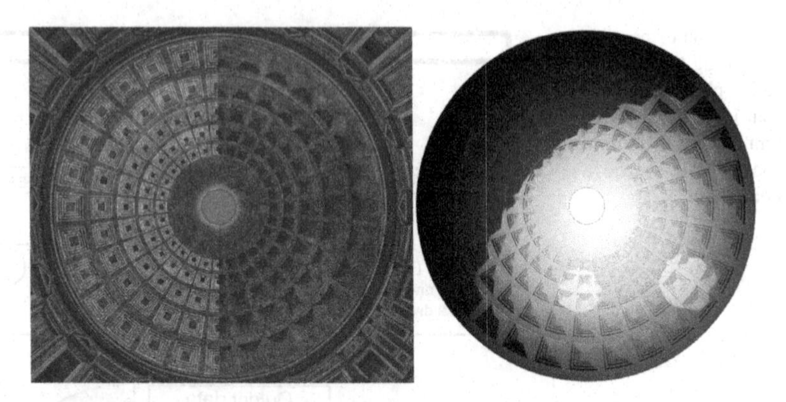

Fig. 9 Pantheon's dome in stereographic projection overlapped by hypothetical design (in *white*). (on the *right*) Stereographic projection of tree-dimensional model generated using the described algorithm. Photo by author

References

1. Chase, S. C. (1993). The use of multiple representations to facilitate design interpretation. In *Proceedings of ARECDAO'93*, Barcelona (pp. 205–217).
2. Gips, J., & Stiny, G. (1972). Shape grammars and the generative specifications of painting and sculpture. In *The best computer papers of 1971* (pp. 125–135). Philadelphia: Auerbach Publishers.
3. Knight, T. W. (1998). Designing a shape grammar. In *Artificial Intelligence in Design'98* (pp. 499–516). Springer.
4. Stiny, G., & Mitchell, W. J. (1978). The palladian grammar. *Environment and Planning B: Planning and Design, 5*(1), 5–18.
5. Stiny, G. (2011). What rule(s) should i use? *Nexus Network Journal, 13*(1), 15–47.
6. Soddu, C., & Colabella, E. (2008). Lumen in Splendore. In *Proceedings of 7th international conference Aplimat*. Bratislava: STU Bratislava.
7. Bartoli, M. T. (1995). Scaenographia vitruviana: il disegno delle volte a lacunari tra rappresentazione e costruzione. In *DISEGNARE IDEE IMMAGINI* (pp. 51–62). Roma: Gangemi Editore S.p.A., n. 9/10.
8. Bartoli, M. T. (1997). *Le ragioni geometriche del segno architettonico*. Firenze: Alinea Editrice s.r.l.
9. Bartoli, M. T. (2010). Il cielo, la terra e le cupole a lacunari. In Mandelli, E. (Ed.), *Disegnare il tempo e l'armonia. Il disegno di Architettura osservatorio nell'Universo, Proceedings of International Congress A.E.D.* Firenze: Alinea Editrice s.r.l.
10. Valenti, G. M. (2009). A computing model for the Pantheon's cupola; from the discrete to the continuous. The ideal continuous model. In G. Grabof, M. Heinzelmann, & M. Wafler (Eds.), *The Pantheon in Rome*. Bern Studies in the History and Philosophy of Science: Bern.
11. Rondelet, G. (1832). *Trattato teorico e pratico dell'arte di edificare, tomo II e III*. Mantova.

Printed in the United States
By Bookmasters